T0186904

1. K.Ya. Kondrat'ev et al. (editors): USSR/USA *Bering Sea Experiment*
2. D.V. Nalivkin: *Hurricanes, Storms and Tornadoes*
3. V.M. Novikov (editor): *Handbook of Fishery Technology*, Volume 1
4. F.G. Martyshev: *Pond Fisheries*
5. R.N. Burukovskii: *Key to Shrimps and Lobsters*
6. V.M. Novikov (editor): *Handbook of Fishery Technology*, Volume 4
7. V.P. Bykov (editor): *Marine Fishes*
8. N.N. Tsvelev: *Grasses of the Soviet Union*
9. L.V. Metlitskii et al.: *Controlled Atmosphere Storage of Fruits*
10. M.A. Glazovskaya: *Soils of the World* (2 volumes)
11. V.G. Kort & V.S. Samoilenko: *Atlantic Hydrophysical Polygon-70*
12. M.A. Margdzhanishvili: *Seismic Design of Frame-panel Buildings and Their Structural Members*
13. E.'A. Sokolenko (editor): *Water and Salt Regimes of Soils: Modeling and Management*
14. A.P. Bocharov: *A Description of Devices Used in the Study of Wind Erosion of Soils*
15. E.S. Artsybashev: *Forest Fires and Their Control*
16. R.Kh. Makasheva: *The Pea*
17. N.G. Kondrashova: *Shipboard Refrigeration and Fish Processing Equipment*
18. S.M. Uspenskii: *Life in High Latitudes*
19. A.V. Rozova: *Biostratigraphic Zoning and Trilobites of the Upper Cambrian and Lower Ordovician of the Northwestern Siberian Platform*
20. N.I. Barkov: *Ice Shelves of Antarctica*
21. V.P. Averkiev: *Shipboard Fish Scouting and Electronavigational Equipment*
22. D.F. Petrov (Editor-in-Chief): *Apomixis and Its Role in Evolution and Breeding*
23. G.A. Mchedlidze: *General Features of the Paleobiological Evolution of Cetacea*
24. M.G. Ravich et al.: *Geological Structure of Mac. Robertson Land (East Antarctica)*
25. L.A. Timokhov (editor): *Dynamics of Ice Cover*
26. K.Ya. Kondrat'ev: *Changes in Global Climate*
27. P.S. Nartov: *Disk Soil-Working Implements*
28. V.L. Kontrimavichus (Editor-in-Chief): *Beringia in the Cenozoic Era*
29. S.V. Nerpin & A.F. Chudnovskii: *Heat and Mass Transfer in tthe Plant-Soil-Air System*
30. T.V. Alekseeva et al.: *Highway Machines*
31. N.I. Klenin et al.: *Agricultural Machines*
32. V.K. Rudnev: *Digging of Soils by Earthmovers with Powered Parts*
33. A.N. Zelenin et al.: *Machines for Moving the Earth*
34. *Systematics, Breeding and Seed Production of Potatoes*
35. D.S. Orlov: *Humus Acids of Soils*
36. M.M. Severnev (editor): *Wear of Agricultural Machine Parts*
37. Kh.A. Khachatryan: *Operation of Soil-Working Implements in Hilly Regions*
38. L.V. Gyachev: *Theory of Surfaces of Plow Bottoms*
39. S.V. Kardashevskii et al.: *Testing of Agricultural Technological Processes*
40. M.A. Sadovskii (editor): *Physics of the Earthquake Focus*
41. I.M. Dolgin: *Climate of Antarctica*
42. V.V. Egorov et al.: *Classification and Diagnostics of Soils of the USSR*
43. V.A. Moshkin: *Castor*
44. E.'T. Sarukhanyan: *Structure and Variability of the Antartic Circumpolar Current*
45. V.A. Shapa (Chief editor): *Biological Plant Protection*
46. A.I. Zakharova: *Estimation of Seismicity Parameters Using a Computer*
47. M.A. Mardzhanishvili & L.M. Mardzhanishvili: *Theoretical and Experimental Analysis of Members of Earthquake-proof Frame-panel Buildings*
48. S.G. Shul'man: *Seismic Pressure of Water on Hydraulic Structures*
49. Yu.A. Ibad-zade: *Movement of Sediments in Open Channels*
50. I.S. Popushoi (Chief editor): *Biological and Chemical Methods of Plant Protection*

CONTINENTAL RIFT FORMATION

AND

ITS PREHISTORY

Continental Rift Formation and Its Prehistory

A.V. RAZVALYAEV

RUSSIAN TRANSLATIONS SERIES

87

1991

A.A. BALKEMA/ROTTERDAM

Translation of: *Kontinental'nii riftogenez i ego predistoriya, Nedra, Moscow 1988*

© 1991 Copyright reserved

Translator   :   Dr. S.P. Ghosh
Technical Editor :   Dr. R. Chakraverty
General Editor   :   Ms. Margaret Majithia

ISBN 90 6191 991 6

Preface

The last 15–20 years have been marked by intensive study of both continental and oceanic rift structures and processing of the results of these investigations towards a new direction in geotectonics, namely, a study of rift formation. Extensive geologic-geophysic investigations in oceans during the fifties and sixties led to the discovery of mid-oceanic rift zones, which stimulated interest in continental rift structures. These studies necessitated reinterpretation of the historic-geologic role of rift formation in the evolution of the Earth.

A new approach towards the relationship between rifts and their pre-rift history was assayed in this study of continental rift formation, i.e., predetermination of inheritance of rift formation from previous history. Apparently rift formation is preceded by a unique tectonic set-up which helps to prepare the lithosphere or 'ripen' it (according to the terminology of E.E. Milanovskii) for rift formation. The majority of workers are inclined to believe that rift formation is not accidental in the course of crustal evolution and that rift structures appear in regions having a specific geologic set-up and history of development. V.E. Khain was the first to indicate the relationship between rift formation and specific belts of late Proterozoic tectonomagmatic activisation subjected to repetitive orogenesis and metamorphic regeneration. V.E. Khain particularly stressed that these belts had served as the arena for continental rift formation in the Phanerozoic. Subsequently, N.A. Bozhko, V. Fyfe, O. Leonardos, E.A. Dolginov, V.M. Moralev and V.P. Ponikarov drew attention to the existence of such belts prior to rift formation.

It has been established that Cenozoic rifts are superposed and, to a large extent, autonomous structures closely related to the earlier developed rift-forming regions [27]. In structural plan, this relationship is expressed in the inherited trend of rifts of old folded zones, faults, fracture zones and other structural characteristics of the substratum, i.e., anisotropy of the rifts. The uniqueness of continental rift formation has been studied well by many scientists, namely, F. Dixie, N.A. Florence, R.V. McConnell, E.E. Milanovskii, A.F. Grachev, N.A. Logachev, S.I. Sherman and others, who analysed the Baikal, East African, Rhine, Rio Grande and other Cenozoic rifts and proved that this phenomenon is fairly widespread.

V.E. Khain, E.E. Milanovskii, N.A. Logachev, N.A. Bozhko, E.A. Dolginov, A.V. Razvalyaev and others have noted a deeper relationship between Cenozoic rifts and Precambrian structures, which maintained their intensive tectonic activity and, consequently, a high thermal regime up to the time of rift formations. Cenozoic and possibly older rifts are usually confined to late Proterozoic fold regions with basite substratum and also to regions with tectonothermal reworking of late Precambrian belts.

However, investigations have shown that under identical geologic conditions, the degrees of manifestation of rift formation differ significantly. The structure and evolution of rift zones and their relationship with the substratum are many times more complex than the known interrelationships; at the same time the causes for a more intimate relationship in rift formations developed at different places, i.e., of destructive tectogenesis, are not yet clear. The problems concerning the prehistory of riftogenesis still remain inadequately studied in geotectonics.

In the present book the author has attempted to establish a theoretical background for the concept of the relationship between continental rift formation and preceding geological history on the basis of the Red Sea and other rift zones. A study of rift formation from the constructive-destructive viewpoint of lithosphere development in combination with the concept of existence of stable endogenetic regimes helped in distinguishing the prerift stage of development of rift zones. The prerift stage includes all the tectonomagmatic processes which were active at some place or the other in rift-forming regions after the formation of a mature continental core and continuing up to the time the structural and morphologic configuration of the rift was achieved. Rift formation preceded specific tectonomagmatic processes, causing specific features of the prerift stage of development of rift zones, such as duration and cyclicity of thermal activation of the mantle as revealed in the generation of magma at different depths in strictly localised linear zones which predetermined the location of future rifts. Continental rift formation is more actively manifested in regions where the lithosphere was heated more and thermally destabilised in the prerift stage. The principle of inheritance, as applicable to rift formation, has been developed from these standpoints and its duality has thus been revealed in its inheritable feature and autonomy. The inheritance is distinguishable at mantle and crustal levels. At the mantle level it was caused by a succession of endogenetic regimes of prerift and rift stages, and at the crustal level by the dependence of rift formation on structural anisotropy of the basement. Such an attempt at studying the continental rift formation corresponds to delineation of the mechanism of earlier stages of its development and roots of geological history.

Presentation of the given problem and its solution stipulates specific conditions in the selection of the objects. Firstly, it is necessary that the substratum of the rift be approachable for in-depth and detailed study; secondly, the prerift activised magmatism should have undergone prolonged development and its area-wise distribution should provide correlation among magmatic complexes both in composition as well as in age.

The Red Sea rift zone overlying the Precambrian basement is a unique region. The processes of generation of endogenetic regimes and successive stages of destruction of the continental crust from the initial stages to the end stages with formation of oceanic type crusts can be traced through its geologic history.

In recent years the study of rift formations has been 'booming' and become 'fashionable' among a specific group of researchers. They often attribute one or the other theoretical concept to the phenomenon with no critical analysis or unduly simplify the rift formation process, thereby denigrating it to merely a morphostructural form. Therefore identification of the characteristics of the composition and evolution of specific rift zones in order to arrive at valid concepts regarding the role of rift formation in the geologic history of the earth is a genuine need.

A large number of specific rift zones must necessarily be studied for purposes of objectivity; some rifts may be of one tectonic type or, in a favourable case, may correspond with a few aspects of rift formation from the point of view of their geological structure. Thus, for example, the Kenyan Rift and, to a lesser extent, the Ethiopian Rift, serve as tectono-types in establishing the evolution of volcanism and its interrelationship with structural stages of later ages. The Red Sea rift zone is a unique object for tracing the link of Cenozoic rift formation with the preceding geologic history, inasmuch as it shows extensive development of prerift magmatism, which is the main indicator of endogenetic regimes. Its excellent outcrops help in tracing the structural features of the rift and its prerift stages of development, an aspect of equal significance.

The solution to this problem lies in a generalised analysis of the formation and geologic history of the Red Sea rift zone for the typical rift stage and the preceding geologic history. Simultaneously, specific attention is given to: 1) delineation of peculiarities of structure and development of the Red Sea rift zone territories during the Precambrian period; 2) formational analysis of magmatic formations as an indicator of endogenetic regimes; 3) distinction of prerift types of endogenetic regimes; 4) structural evolution of rifts depending on the character of prerift endogenetic regimes, and 5) deep-seated geodynamics and destructive mechanism of the continental crust in active continental rift forming regions.

The present book is the result of many years of investigations by the author in Syria, Egypt and Sudan. In addition to original data, voluminous geologic-geophysical information on rift formation has been summarised, primarily pertaining to the Red Sea, Aden and Western Arabian rift zone, compiled from the publications of V.E. Khain, E.E. Milanovskii, A.F. Grachev, N.A. Logachev, A.I. Almukhamedov and others. Data were also collected from the publications of international symposia on rift formation, held in Irkutsk (1975), Stuttgart (1975–76), Oslo (1977), etc., and also from numerous articles by Soviet and non-Soviet geologists.

The results of numerous investigations carried out by Soviet geologists under the guidance of V.V. Belousov on rift zones in Eastern Africa are of great significance. The author considers his study of the Red Sea rift zone an extension of these investigations at the northern extremity of the African-Arabian rift belt.

Insofar as the role of rifting in the early stages of the earth's evolution (Archean and early Proterozoic) has already been described by E.E. Milanovskii in *Riftogenez v istorii zemli* [Rifting in the Earth's History], it is advisable to examine the prehistory of continental rifting in the Phanerozoic period and to pay specific attention to the theoretical foundation of the prerift stage and to the predetermined continental rifting concept.

Various aspects of the problems of rift formation as a whole and of the African-Arabian rift belt in particular are discussed in the publications of N.A. Logachev, E.A. Dolginov, E.N. Isaev, N.A. Bozhko, I.V. Davidenko, V.G. Lazrenkov, L.E. Levin, V.G. Kaz'min, V.N. Moskaleva, V.T. Ponikarov and others.

While studying the problems of rift genesis, the main results were discussed with scientists who had conducted investigations in Syria, Egypt and Sudan, namely, V.V. Balkhanov, A.N. Vishnevskii, A.I. Krivtsov, I.A. Mikhailov, E.D. Sulidi-Kondrat'ev, G.P. Shakhov and others. Similar discussions were also held with specialists in the National Geological Surveys of these countries. Valuable critical comments were offered by V.E. Khain, E.E. Milanovskii and A.F. Grachev while preparing this book for publication.

Contents

1

Peculiarities of Tectonic Development of the Arabian-Nubian Shield in the Precambrian

(7) The probable dependence of young rift formation on the structural anisotropy of the basement, i.e., 'frame' of rift genesis came to light through the studies of N.A. Florensov, E.E. Milanovskii, A.F. Grachev, N.A. Logachev, S.I. Sherman, A.V. Goryachev, F. Dixie and R.B. McConnell. The role of the substrata in rift genesis was analysed from this viewpoint based on the examples of the East African, Baikal, Rhine and other rift zones and belts. However, one enormous rift zone of the African-Arabian rift belt was not included in this sphere of analysis, namely, the Red Sea rift zone. Yet mainly this zone, which extends for not less than 2000 km and is confined within the Precambrian 'frame', conceals in itself many as yet unknown forms of interrelationship among rifts such as 'frame' or 'frame'—rift—ocean.

Until recently, the problem of interrelationship of rift genesis with the substrata remained inadequately studied for the Red Sea rift zone and had not been practically reviewed in available literature. Only evidence of the presence of Precambrian dykes located parallel to the Aquaba Gulf rift exists and the relationship of the Red Sea rift with late Proterozoic fold belts has been postulated on general considerations. Concepts on the Precambrian development of this territory are schematic, on the whole, which is explained by the absence of adequate study of the Precambrian and lack of any reliable correlational scheme.

Main Stratigraphic and Tectonic Features of the Arabian-Nubian Shield during the Precambrian

Contemporary concepts . about the stratigraphy and tectonism of the Arabian-Nubian shield during the Precambrian are based on investigations conducted by geologists of other countries, e.g., I.G. Gass, J.R. Vail, A. Kröner, I. Delfour, D.C. Almond and others and also by

Soviet geologists, such as E.A. Dolginov, A.V. Razvalyaev, V.G. Kaz'min, V.N. Kozerenko, V.S. Lartsev, etc. The problems of stratigraphy and tectonics of the Red Sea rift zone during the Precambrian have been reviewed to some extent in the publications of V.E. Khain, L.I. Salopa, S.E. Kolotukhina and others in the context of the entire African continent or its individual parts.

Considering that the Precambrian of the Arabian-Nubian shield has been inadequately studied geologically and no reliable correlational scheme is available, special attention has been paid in this chapter to the Red Sea mountains in Egypt and Sudan, studied by the author, and the history of the Red Sea rift zone in the Precambrian. This is necessary for understanding the interrelationships of rift genesis with the Precambrian substrata.

Three structural-compositional complexes have been distinguished (8) in the Precambrian of the Arabian-Nubian shield based on composition, degree of metamorphism and structural characteristics: the lower one—amphibolite schist and gneiss, locally attaining granulite facies of metamorphism (Katarchean-lower Proterozoic); middle one—green schist facies metamorphism of volcano-sedimentary rocks (lower-middle Riphean); and upper one—molasse sediments with initial stage of metamorphism (upper Riphean-Vendian).

The older Katarchean-lower Archean gneissic and schistose complex stands out structurally as the 'basement complex' on which the geosynclinal troughs were formed during the late Proterozoic period. This complex is extensively developed in the Nubian part of the shield and shows a gradual transition towards the Central African craton (Figs. 1 and 2). In the remaining larger part of the shield, it comprises the Central massif type larger blocks bordered by late Proterozoic fold belts. The block boundaries are often tectonic in nature and traversed by large-scale faults. The schistose-gneissic complex, occurring over extensive areas of Central Sudan, Southwestern Ethiopia and partly in Saudi Arabia, shows the same structure and is represented almost exclusively by biotitic and amphibolitic paragneisses and locally by leucocratic gneisses with granulitic texture. The gneisses are migmatised to varying degrees. Besides gneisses, the complex shows extensive occurrence of amphibolite-, biotite-, biotite- muscovite-, quartz-sericite and quartz-chlorite-schists, quartzites, marbles and calc-silicate rocks.

Notwithstanding the fact that our knowledge about the Precambrian geology of the Arabian-Nubian shield has been considerably enriched in recent years, the problem regarding the time and mechanism of formation of the continental crust still remains a debatable issue. The problem of segregating the gneissic complex as an independent structural subdivision and, consequently, the problem of tectonic type of basement for

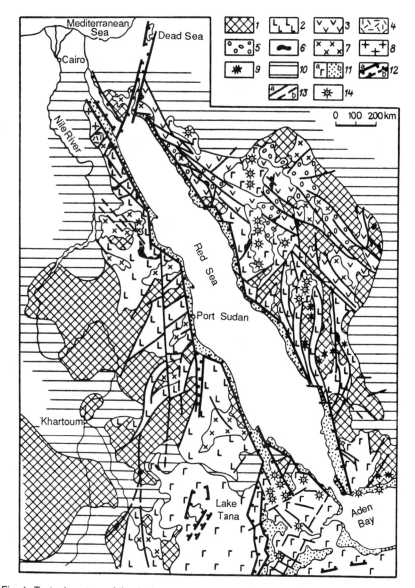

(9) Fig. 1. Tectonic set-up of the Arabian-Nubian shield.

1—Katarchean (?)-Archean granite-gneissic basement (amphibolitic with relicts of granulite facies metamorphism); 2–3—green schist volcanogenic-sedimentary complexes of the Red Sea folded region (2—early-middle Riphean, 3—late Riphean); 4–5—orogenic complexes (4—late Riphean, 5—Vendian-early Palaeozoic); 6—early-middle Riphean Alpine type ultrabasite complex; 7—late Riphean syntectonic calc-alkaline granites (batholithic complex); 8—late Riphean-Vendian-early Palaeozoic (?) post-tectonic subalkaline and alkaline granites; 9—late Riphean-Vendian-early Palaeozoic 'layered' gabbro intrusives; 10—platform cover; 11—riftogenic complexes (a—volcanogenic, b—sedimentary); 12—faults in rift depression (a—established, b—assumed); 13—regional faults (a—established, b—assumed); 14—volcanoes.

the late Proterozoic fold belts, have remained unsolved insofar as the interrelationship of gneisses with green schist rocks is concerned. The gneisses are considered either old sialic substrata on which late Proterozoic intracratonic troughs were formed, or as metamorphic analogues of volcanogenic-sedimentary rocks filling these troughs, formed in the oceanic crust. The approach in establishing either point of view depends on the time of formation of the granite-metamorphic band of the Arabian-Nubian shield, the nature of volcanogenic-sedimentary troughs and the dynamics of their development together with the associated faults at the mantle level (ultrabasite furrows).

The problem concerning partitioning of the ancient gneissic complex of the Arabian-Nubian shield is further complicated by the fact that the ancient gneissic blocks underwent remobilisation and thermal effect in the late Riphean-early Palaeozoic period of the Pan-African orogeny, leading to radiological regeneration of rocks—a phenomenon widely prevalent in the late Precambrians of Africa. Data on the detailed geological mapping (11) of the western boundary of the Red Sea show that intensive and repeated remobilisation of the gneissic complex led to the formation of fold structures in the marginal part of the gneissic blocks. These structures are conformable to those bordering Proterozoic troughs of green-schist rocks. This significantly illuminates the primary interrelationship between the gneisses and green-schist rocks.

Data from the latest researches of the author in the Red Sea mountains of Sudan assume specific significance in solving the controversial distinction of the oldest schistose and gneissic complexes. It has been established that the schistose and gneissic complexes comprise two separate tectonic blocks in this region (see Fig. 2). One is the Derudeb block situated in the southern part of the Red Sea mountains, the second one is the Kashebib block, located in the north. B. Rukston had earlier distinguished the schistose and gneissic complexes as a 'primitive system' whereas M. Kabish identified them as the Kashebib series.

The schistose and gneissic complex of the Derudeb block can be subdivided into three parts according to the degree of metamorphism and structural peculiarities: ultrametamorphic (the Imas gneiss), gneissose

---

(10) Fig. 2. Schematic depiction of tectonic zoning in the Precambrian Arabian-Nubian shield.

1—pre-upper Proterozoic granite-gneissic basement (a—surficial, b—under platform cover); 2—late Proterozoic Red Sea folded region; 3—late Proterozoic faults; 4—Cenozoic riftogenic faults; 5—upper Proterozoic ultrabasites; 6—ancient granite-gneissic blocks of the Sudanese Red Sea Mountains: I—Derudeb, II—Kashebib; 7—late Proterozoic folded belts: 1—Sudan-Arabian, 2—East African, 3—East Egyptian, 4—Eritrean, 5—South Arabian.

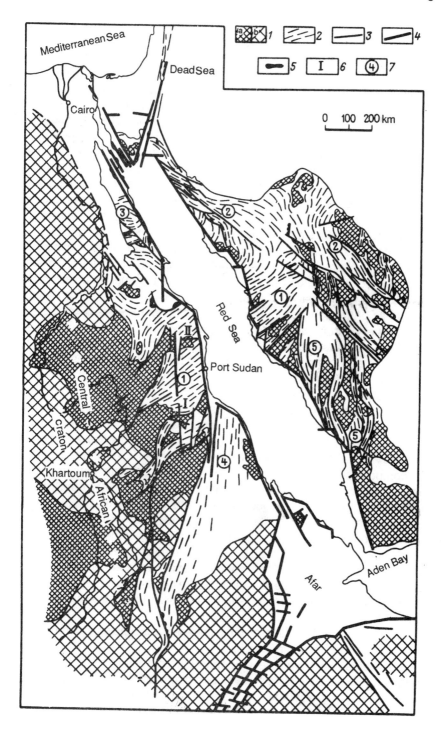

and schistose. The older part of the complex is represented by products of ultrametamorphism, namely, by banded biotitic gneisses, 'shaded' migmatites and anatectic granites. Various types of gneisses and quartzites are of subsidiary significance. The anatectic granites, with distinct porphyroblastic texture at places, show a gradual transition into the migmatites. The ultrametamorphic rocks constitute a block small in area with a predominantly sublatitudinal strike and isometric cupola-type folds. That these rocks have a tectonic relationship with the younger complexes is revealed by the thick zones of schistosity, mineralisation and dyke systems. Amphiboles and amphibolites, among which thin layers of biotitic gneisses are also observed, predominate at the base of the section in the gneissic part. The biotite gneisses dominate in the upper sections, where discordant lenses of amphibolites, apparently representing metamorphosed basic volcanic rocks, are common. The gneissic part of the complex is 2 to 3 km thick. The schistose part of the section in the complex is characterised by alternating amphibolitic, chloritic, chlorite-amphibolitic, quartz-sericitic and other schists together with rare marble layers. The thickness of the schistose rocks is 1.8 km. The average thickness of the schistose and gneissic complex exceeds 5 km.

Garnet-bearing rock types are developed both in the gneissic as well as in the schistose parts of the complex. The mineralogical paragenetic relation of such complexes indicates that the original composition of the rocks was sedimentary or sedimentary and volcanic, which then meta-morphosed to amphibolite facies.

(12)   The other section of the schistose and gneissic complex (Kashebib series) was studied at the western boundary of the Red Sea at 21° N lat. Here, as in the Derudeb block section, the schistose and gneissic complex is subdivided into a lower band which is essentially gneissic in composition, and an upper band that is schistose in composition. The gneissic band consists predominantly of about 500 m thick biotite-hornblende gneisses. The schistose band starts with a light-coloured quartz-feldspar schist layer of 130 m thickness and quartzites with intercalations of muscovite schists. The main part of the band consists of amphibolite schists and amphibolites; micaceous schists occur only to a limited extent. Bands and horizons of massive and banded white marble, up to several tens of metres thick and extending for a few kilometres along the strike, are characteristic for these strata. The schistose strata are 1000 m thick.

The average thickness of the schistose and gneissic complexes in the section under examination exceeds 1000 m.

Garnet is extensively developed in both groups of rocks, particularly in the upper one. Extensively developed pegmatitic bodies, occurring both concordantly and discordantly along schistosity or gneissosity and also intensively dislocated pegmatitic injections are characteristic for the complex.

The schistose and gneissic complexes are completely dislocated and their structures reveal overturned folds. The predominant strikes of the folds vary from submeridional to north-easterly. The complex is intruded by syntectonic gneissose granites and their emplacement is controlled by the general strike of the folded structure.

The interrelationship of the schistose and gneissic complex with the younger volcanogenic sedimentary Nafirdeib series of rocks (lower to middle Riphean) in which metamorphism is confined within the epidote-chlorite grade has been a topic of much discussion. The dual approach in dealing with the nature of schistose and gneissic complex in the earlier mentioned aspects is partly explained by the fact that, for the most part, the contact of these series is tectonic and is accompanied by an intensive mylonitisation zone together with the development of diaphthorite schists. The findings at the contact of these complexes in the Habeit-Main region are very significant. Here, it was thought earlier, the real interrelationship of the complexes has been concealed by intensive mylonitisation and schist formation with associated mineralogical changes due to retrograde metamorphism. The author's detailed study in this region helped to reveal an anticlinal fold with schistose and gneissic complexes of rocks occurring in its core. The strike of the anticlinal axis is meridional. The fold axis plunges in a northern direction. Amphibolite schists gently plunge below the lower and middle Riphean Nafirdeib pyroxene-andesite porphyry series of rocks, which are practically unaffected by schist-forming processes and mineralisation in the periclinal closure of the fold. The contact of crystalline schists and andesites is fairly sharp; occasionally an occurrence of basal conglomerate is noticed. The foregoing interrelationships leave no doubt about the lower stratigraphic position of the schistose and gneissic complex.

(13)   Petrographic studies showed that rocks of different metamorphic grades are in contact with each other, e.g., crystalline amphibolitic schists altering into minerals of garnet-almandine facies and volcanic rocks with traces of epidote-chlorite-sericite (not higher than green schist) facies of metamorphism. It is noteworthy that the characteristic occurrence of pegmatites in the schistose and gneissic complexes is absent at the contact of younger volcanites of the Nafirdeib series.

The stratigraphic position, garnet-amphibolitic grade of metamorphism, intense folding, ptygmatisation of the schistose and gneissic complexes and their relationship with the pegmatites—all these phenomena underline its structural partitioning and prompt us to consider it the ancient basement of the Arabian-Nubian shield. A comparison of detailed studies on two sections of the schistose and gneissic complexes, separated from each other by not less than 400 km, revealed compositional uniformity, indicating thereby the considerable homogeneity of the palaeotectonic conditions of their formation.

The Khamis-Mushkhayat complex represented by gneisses and amphibolites and developed in the form of bands and 'shadow relics' among wide granite gneissic outcrops are considered the ancient basement for the late Proterozoic fold belts in the Arabian peninsula and Saudi Arabia. According to R.G. Kolman's data, the andesite rock series of the lower and middle Riphean were unconformably deposited on these gneisses, often with conglomerates at the basement.

Ultrametamorphic processes leading to the formation of migmatites, development of anatectic granites, the sublatitudinal strike of the structural components, similar to the latitudinal strike of the oldest granite gneissic complexes of the Tanganyikan shield and the Mozambique belt, prompted a comparison of the detected schistose and gneissic complexes with the Katarchean formations of the Tanganyikan, Central African and other cratons. According to position in the geological section, petrographic composition and degree of metamorphism, they are parallel to the schistose and gneissic strata extensively developed in north-eastern Africa. Analogues of these formations have been detected in northern Sudan, Libya, Ethiopia, Somali and Saudi Arabia (Table 1).

Thus, the oldest schistose and gneissic complexes, formed in the Katarchean-early Archean (?) period, are found to occur over an extensive territory of the Arabian-Nubian shield. These are represented by the grey gneisses of Sudan, variegated gneisses, charnockites and other granulite facies rocks of Libya and Ethiopia and Vatian, Karamuja and Kittam series of Uganda and other places. On the whole, a higher degree of metamorphism is characteristic for the enumerated series of rocks. They are separated from overlying schists and gneisses in Sudan by an angular unconformity. Studies by J.R. Vail and D.C. Almond showed that granulite facies rocks are developed in the Sudanese territory on a much wider scale than was earlier presumed. The granulite-containing complexes show a transition from here into the Central African craton (14) where the Western Nile Series of rocks might be the possible analogues. Consequently, it may be assumed that the lower horizons of the schistose and gneissic complexes of the Arabian-Nubian shield in Sudan, Libya and Ethiopia correspond to the Katarchean in age. For schistose and gneissic complexes of other regions of the Arabian-Nubian shield, a Katarchean age is less definite. However, it may be presumed that rocks of this age are more extensively developed in the areas of the shield and these entered into the composition of the vast Central African craton in the early stages of its development (Fig. 3).

The presence of the ancient Archean-early Proterozoic (?) basement

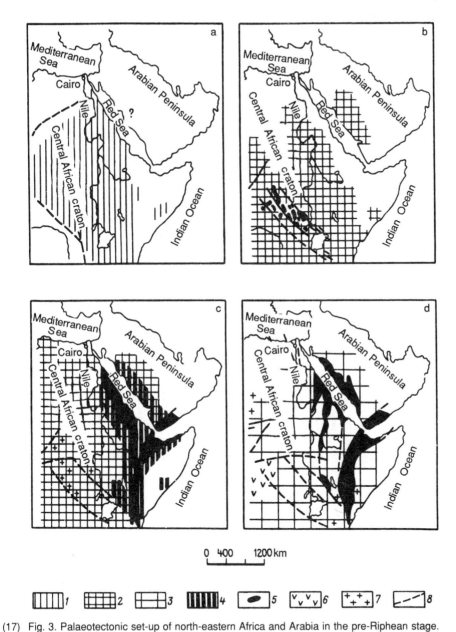

(17) Fig. 3. Palaeotectonic set-up of north-eastern Africa and Arabia in the pre-Riphean stage.

a—Katarchean, b—early Archean; c—late Archean; d—early Proterozoic; 1—primary crust formed in the Katarchean (?); 2—primary crust which underwent granitisation in the early Archean; 3—stable blocks of the continental core; 4—intracratonic troughs of the miogeosynclinal type; 5—same, eugeosynclinal type; 6—volcanogenic-sedimentary cover; 7—orogenic granitoid magmatism; 8—assumed faults.

(14—15)

Table 1. Correlational scheme for the early Precambrian of north-east Africa and Arabia.

Age	Libya (Uwaynat hills) (G. Klerkx [41])	Sudan		
		Bayud Desert (K.D. Meinhold [45], with changes by the author)	Nile valley, Sobat River region (D. Almond [35])	Red Sea hills (A.V. Razvalyaev)
Upper Archean-Lower Proterozoic (?)	*Ain-Dua series* Granite-gneisses, at places migmatised, anatectic granite intrusions with amphibolite layers; 1836 million years from anatectic granites	Bayud Formation *Absol series* Kyanite-garnet-staurolite schists with amphibolite layers and lenses of marble and siliceous rocks *Kurmut series* Amphibolite facies metasediments accumulated under shelf conditions; quartzites, quartzitic schists, red quartz-feldspathic gneisses, biotite and biotite-hornblende gneisses; micaceous, graphitic schists, calc-silicate rocks with metavolcanite intercalations	*Metasedimentary series of rocks of amphibolite facies metamorphism* Quartz-feldspathic garnet gneisses, amphibolites, marble and quartzite lenses	Kashebib Series Amphibolites, garnet-amphibole, quartz-feldspar and mica schists Biotite-amphibole, garnet-bearing gneisses, migmatites, quartzites, marbles
Katarchean-Lower Archean (?)	*Kurkur-Murr series* Biotite gneisses, charnockites containing metaquartzites and calc-silicate rocks (2673 million years, Rb-Sr method) of granulite facies metamorphism	*Rahaba series* Acidic quartz-feldspar gneisses, biotite-garnet gneisses and amphibolites, migmatised and granitised rocks	*'Grey gneiss' complex* Biotite-gneisses, migmatites, amphibolites with relict rocks of mineral association of granulite facies metamorphism	*Imas ultrametamorphic complex* Granite-gneisses, migmatites, anatectic granites

Table 1 (Contd.)

Age	Saudi Arabia (J. Delfour [39])	Ethiopia (V.G. Kaz'min and others)	Somali (M.N. Kozernko, V.S. Lartser)	Northern Uganda, (D. Almond [35])
	Abu-Harik series—grey gneiss complex Biotite gneisses, highly migmatised amphibolites partially granitised			
Upper Archean-Lower Proterozoic (?)	Adzal formation Crystalline schists amphibolites quartzites	Adola group Amphibolites: amphibole, chlorite, talc schists; quartzites and graphite-bearing rocks. Wadera group proto-Platform cover) Meta-sandstones, quartz-feldspar rocks, mica schists	Borama Laferug suite (upper part) Biotite-amphibole shists with garnet, gneisses, orthoamphibolites (lower part) Biotite-amphibolites quartz-feldspar rocks	Karasuk series (Turoka) Gneisses, amphibolites, marble calc-silicate rocks
Katarchean-Lower Archean (?)	Ancient basement gneisses Hamis-Mushhayat Granite gneissis, migmatites	Arero group Biotitic and biotite-amphibole gneisses, pyroxene, pyroxene-garnet gneisses, quartz-silicate rocks, quartzites; relicts of granulite facies metamorphism	Olontole suite Biotite-amphibole gneisses, amphibolites	Kitgam group (2900 million years) Biotitic and biotite-amphibole gneisses; quartz-feldspar rocks, granulites

in Arabia was doubted until recently, particularly in publications outside the USSR. Such a postulate is based on theoretical premises regarding the existence of an extensive ocean in the Arabian-Nubian shield in the late Proterozoic. Accordingly, formation of the continental crust took place in the late Proterozoic through the emergence of a series of island arcs in the north-east adjacent to the Benioff Zone, which was subsiding to the west (or east). At this stage, the island arcs were laid on the primary oceanic crust, which underwent subsequent compression and piling up, and as a result closure (shutting up) of the oceanic basins and accretion of regenerated continental crust took place. From such a viewpoint, the gneissic strata are regarded as granite-gneissic cupolas formed under collision and represent metamorphic analogues of late Proterozoic green-schist rocks.

Based on the factual problem of the gneissic basement of Arabia in the international project for the correlation of the Precambrian Arabian-Nubian shield, attempts were made to determine the nature of the Saudi Arabian gneisses in areas where they are considered to be the ancient sialic basement on the basis of structural, stratigraphic and metamorphic interrelationships. Data on the Rb/Sr and $^{87}Sr/^{86}Sr$ ratio obtained from migmatites and granite-gneisses from two regions (Dahul and Jugjug) showed that in one case (Dahul) the gneisses developed from syntectonic granites of Pan-African age (600 million years) and in the second (Jugjug) case, represent intrusives in some still unknown ancient strata injected during an interval of 870–780 million years.

The latest investigations have confirmed the presence of components of ancient pre-late Proterozoic continental crust in the structure of the Arabian-Nubian shield, and, what is important, in its Arabian part. A publication on age determination of rocks (U-Pb, Rb-Sr and Sm-Nd methods) in eastern parts of Saudi Arabia [47] is more indicative in this regard. These investigations established that notwithstanding the rocks having undergone complex thermal evolution, evidence for the existence of the early Proterozoic (1630 million years) continental crust is available; lead isotope data do not exclude an even earlier (late Archean) age. Data on tectonic aspects have been published by D.B. Stoeser, V.E. Camp [48] and J.R. Vail [49]. In these publications, the Arabian-Nubian shield structures are considered a combination of ancient micro-continental plates, separated by folded late Proterozoic island-arc type collision belts or sutures. J.R. Vail [49] considers that starting from the boundary of 2600 million years, the Arabian-Nubian shield constituted a part of the Gondwana, which later underwent (up to 950 million years) fragmentation due to rifting. From this viewpoint, the continental blocks in Arabia were formed during 715–630 million years by fragmentation of the Gondwana blocks and formed the Arabian Neocraton.

Thus, at present there is no basis for denying the presence of Archean granitic and metamorphic layers in the structure of Arabia. French geologists have recently distinguished a gneissose migmatitic complex in southern Arabia (Yemen People's Democratic Republic). This may be correlated with the Archean formation of eastern Africa in degree of metamorphism, structural relation with younger Precambrian formations and formational composition. These data indirectly attest to the Archean age of the granitic and metamorphic strata in Arabia. Moreover, granulite facies metamorphism has been observed in isolated cases in the composition of the latter. This phenomenon may indicate the metamorphic and structural isolation of the Arabian protometamorphic strata and its Archean age, which is confirmed by radiologic data.

The north-eastern African territory entered into a stage of intensive regeneration of the oldest protocontinental crust from the beginning of the late Proterozoic. The thick strata of volcanic-sedimentary rocks of green-schist facies metamorphism were formed at this stage (Fig. 4). The late Proterozoic complex shows a complex composition, comprising a number of lithologic-stratigraphic subdivisions or formations according to the concept of geologists outside the USSR: the series or formation Abu-Ziran and Atala in Egypt; Odi in Nafrideib and Butan schists in Sudan; Beish, Bakha, Jida and Khalaban in Saudi Arabia; Tzaliet in Ethiopia; Tha-Lab and Garish in the Yemen People's Democratic Republic and Abdul Kard Heis in Somali (Table 2).

The significant characteristic of the green-schist complex is its volcanogenic composition. The extrusive rocks are represented by andesites, basalts, dacites and rhyolites, altered to green schist (andesite and basalt porphyries, rhyolite and dacite porphyry). Agglomeratic and amygdaloidal lavas and also pillow lavas are found among effusive rocks. Extrusive* rocks of the volcanic series are represented by agglomerates and tuffs including coarse-grained varieties of the latter. Tuffconglomerates, tuff-breccia and tuffaceous sandstones are associated with tuffs. More metamorphosed varieties, such as mica-bearing crystalline schists, rarely gneisses and also marbles and conglomerates, also occur in the complex. The common types of mica schists are chlorite-sericite, (22) quartz-sericite, quartz-graphite, rarely, amphibole and biotite-muscovite, and amphibole-gneiss rocks which were possibly formed by contact metamorphism at the contact of batholythic granites exposed in the core of the anticlinoria.

The late Proterozoic green-schist complex has been well studied in Egypt and Sudan.

The Nafirdeib series comprises two members in the type stratigraphic

---

*Both the terms 'extrusive' and 'effusive' are used in the Russian text—Translator.

Table 2. Correlational scheme for the early Precambrian Arabian-Nubian shield.

Age, million years	Egypt	North-east Sudan	Saudi Arabia		Ethiopia	Somali	Yemen Democratic People's Republic
			South	North			
Vendian	Rhyolites of NE desert of Egypt	—	Jubala Formation	Shammor Formation	Mateos Formation	Inda-Ad Formation	
			Shammor Rhyolite	Fatima Formation	Shiraro Formation		
	Hammamat Series	Avat Series	Murdama Formation	Murdama Formation	Didakama Group		
670	Saramuzh conglomerates of Jordan		Halaban Formation	Hulafa Series			?
Upper Riphean		Asotriba Volcanites, Homager Series	Urd Group		Tambien Group		Habar Group
		Dohan Series	Alba Formation				
1000							

(contd.)

Table 2 (contd.)

	Abu-Ziran Formation; Shale-geywacke Series	Nafirdeib Series (upper part); Nafirdeib Series (lower part)	Jida Formation; Baha Formation; Beish Formation	Abt Formation; Arridania Formation; Ophiolite Complex	Zaliet Group; Marmor Group		Harish Group, Tha-Lab
Lower-Middle Riphean							
1650 ± 50	Migif-Harafit Series	Kashebib Series	—	Ajal Series	Adola Group	Abdul Kard-Heis Suite	Aden metamorphic group (Akhbar)
Katarchean-Lower Proterozoic (?)	—	Imasa-ultra metamorphic complex; Bayud series gneisses	Hamis-Mushhyat Gneisses	Ancient basement	Wadera Group; Aero Group	Borama Laferug Suite (upper part); Borama Laferug Suite (lower part); Olontole Suite	Ancient gneisses

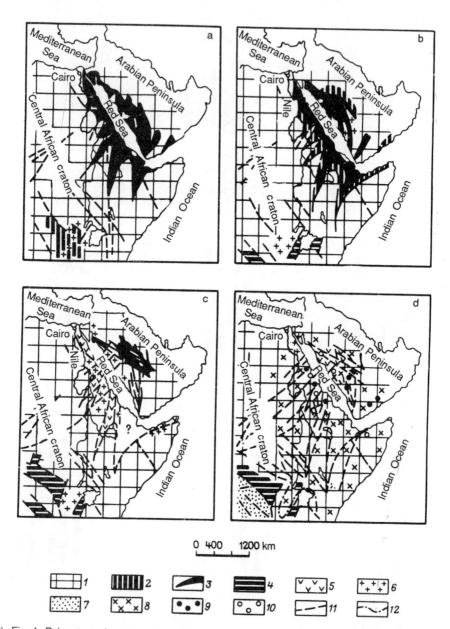

(19) Fig. 4. Palaeotectonic set-up of north-eastern Africa and Arabia during Riphean-Vendian stage.

a—early Riphean; b—middle Riphean, c—late Riphean, d—Vendian; 1—stable continental crustal blocks; 2—intracratonic (rift-form) miogeosynclinal type troughs; 3—intracratonic eugeosynclinal troughs; 4—palaeoavlacogens; 5—orogenic volcano-plutonic troughs; 6—granitoid magmatism and orogenic uplifts; 7—plate complexes; 8—region of Pan-African activisation; 9—manifestation of activised basite and alkali-basite magmatism; 10—same, alkaline-granitoid magmatism; 11—assumed line of faults, 12—provisional boundary of distribution of Pan-African activisation.

section of Sudan [29]. Its lower part is represented by a thick volcanogenic stratum and the upper part is sedimentary. The volcanogenic rocks are predominantly andesites, andesite-basalts and their pyroclasts. Pyroxene and plagioclase porphyries play a significant role among the andesites. Amygdaloidal basalts have an insignificant role in the volcanic rock sections. Acid volcanic rocks (rhyolites) have a limited occurrence in the series. Rhyolite bodies of considerable volume and area, earlier included in the Nafirdeib series, are actually a younger extrusive-effusive complex of acidic rocks comparable with the acid volcanic rocks 'Imper porphyry' of the Dokkhan series occurring extensively in the Red Sea mountains of Egypt. This does not preclude the presence of acid volcanic rocks in the composition of the series. However, the former are not so extensive and readily distinguishable in their structural and geological position, colour and texture. These are mainly rhyolites and rhyolite-dacites forming insignificant sheets, mainly in the central part of the volcanic rocks of the series.

The upper part of the Nafirdeib series consists of interbedded sandstones, greywacke, conglomerates and marbles. Clastic material of local significance, essentially terrigenous in composition, and the character of interlayering with components of weak rhythmicity impart a coarse flysch appearance to these rocks. The thickness of the volcanogenic rocks is 8 km and of the sedimentary rocks is 4 km. The total thickness of the Nafirdeib series exceeds 12 km.

The older part of the section of upper Proterozoic green-schist complex of Saudi Arabia is represented by the Beish, Bakha and Jida formations and is analogous to the Nafirdeib series of Sudan. According to the data of D. Schmidt and colleagues, the formations comprise greywacke, silica, graphite schist, marbles and intraformational conglomerates. Basalts and andesites predominate among volcanic rocks, while dacites and rhyolites and their pyroclastic equivalents, agglomeratic and amygdaloidal lavas and also pillow lavas are rarely seen. The extrusive rocks are represented by tuffs, agglomerates and volcanic bombs. Tuff conglomerates and tuff breccias are common.

The Abla and Khalaban formations in Saudi Arabia, unconformably lying one above the other, belong to younger rocks of the green-schist facies. The Abla formation consists of a basal boulder conglomerate, arkosic sandstones, siltstones and marbles. The Khalaban formation is predominantly volcanogenic and is andesite-basalt in composition. Basal and intraformational conglomerates, greywackes, siltstones, argillites and marbles have limited significance in their composition. Fragments of intrusive rocks are common among the clastic components. The volcanic rocks are represented by varieties ranging from andesites up to rhyolites; however, the former predominates. The thickness of the Khalaban

(23)

formation is estimated to be 12 to 15 km and, according to D. Schmidt, exceeds 21 km in the south-east of Saudi Arabia.

Data from petrochemical investigations of the volcanic rocks of the Saudi Arabian green-schist complex show that volcanism underwent progressive evolution from a weakly differentiated tholeiite stage (Beish, Bakha and Jida formations) to the calc-alkaline stage (Khalaban).

The age of the Nafirdeib series and its enumerated analogues has long been considered to be late Riphean. However, data available in recent years attest to a younger age. Thus, the radiological age determination of the green-schist rocks in Egypt done by El. Shazli placed them between 1195–1293 million years. M. Yu. Manucci considered these values not so much representative of the age of the rocks per se but rather of the regional metamorphic epoch. According to the data of R. Colman and R. Flex and colleagues, the age of the 'bedded' gabbro intrusives penetrating the monotonous volcanic-sedimentary complex in Saudi Arabia is in the order of 470–500 to 1374 million years (K-Ar method). The age of granites and granodiorites penetrating and partially co-folded with the volcanogenic-sedimentary complex in Egypt, Sudan and southern Arabia has been established as 950–1000 million years (K-Ar and Uranium methods). Thus, inclusion of the green-schist complex in the Kibar tectonomagmatic cycle may presently be considered sufficiently reliable. Moreover, these data do not exclude an older age of the rocks. In the opinion of G.A. Schuber (personal communication), their age may be Karelian. A younger age of the green-schist complex has been established in Saudi Arabia as one moves north and north-east. Thus, the age of the extensively developed Khalaban series here is accepted as late Riphean at this place. The andesites of this series are intruded by calc-alkaline granites 775–725 million years old (Rb-Sr method). Age dating of the andesites themselves shows their age as 761 ± 4 and 775 million years.

The upper Riphean-Vendian complex comprises weakly metamorphosed terrigenous and volcanogenic sediments deposited over older folded Precambrian rocks with sharp angular unconformity. Weak dislocational features are characteristic of them. It (the complex) mainly fills up narrow faulted graben-synclinal trough structures having a gentle dip of constituent rocks. The degree of dislocation of the rocks increases only in the faulted and extremal parts of these troughs. The complex under discussion includes the Dokkhan and Khammamat series in Egypt, Asotriba volcanites and Avat series in Sudan, the Didikama (?) and Shiraro groups in Ethiopia, the Inda-Ad formation in Somali, the Ghabar group in southern Yemen, the Fatima, Shammor, Jubala and Murdama formations in Saudi Arabia and the Saramuzh conglomerates in the Sinai Peninsula (see Table 2). Siliceous and phyllitic schists, greywacke, arkosic and quart-

(24) zose sandstones, limestones, dolomites, acidic and rarely basic volcanic rocks play a significant role in the composition of this series.

In Saudi Arabia, only the Fatima, Shammor and Jubala formations, of limited occurrence mainly in the northern part of the Arabian shield, were earlier included within the molasse complex. The Fatima and Jubala formations consist of basal boulder conglomerates, fine and coarse-grained arkosic sandstones and siltstones. Bands of clay, siliceous limestones and dolomites occur to a limited extent. The essentially volcanogenic Shammor formation includes sheets of rhyolites, rhyolitic ignimbrites and andesite-basalt sheets. Terrigenous deposits of the Murdama formation, represented by basal and intraformational conglomerates, arkoses, siltstones and marbles are presently considered molasse formations. These factors significantly increase the role of molasse formations in the structure of the Arabian shield.

The sedimentary-volcanic complexes of the cratonisation stage in the Arabian-Nubian shield are penetrated by numerous discordant instrusions of alkaline and subalkaline granites, known in geological literature as 'young', 'rose' or 'red' granites and as the 'gattar' granites in Egypt. The latter are 640—450 million years old according to K-Ar and Rb-Sr methods.

Tectonic Nature and Role of Ultrabasite Belts in the Structure of the Arabian-Nubian Shield

The ultrabasite complexes occurring along the late Proterozoic deep-seated faults, i.e., ultrabasite furrows or belts, are significant in the Precambrian structure of the Arabian-Nubian shield. Any analysis of the development of the Precambrian Arabian-Nubian shield would be incomplete without a review of these structures.

Late Proterozoic ultrabasites have long been known to occur in the Arabian-Nubian shield. They border the Red Sea in the form of isolated massifs of fragmented and discontinued zones (Fig. 5). However, these ultrabasites have become the object of thorough study only in recent years. A number of publications have appeared on the problems of the Arabian-Nubian shield ophiolites, reviewing either the composition of the individual massifs or the tectonic features of the belt as a whole. It should be noted that segregation of the ophiolitic complexes is not always convincingly established in these publications and is more often presented as a simple postulation. The petrochemical characteristics of the ultrabasites are cited only for the greywacke, Jebel-Vask massifs in the Arabian peninsula by A. Baker and a few others. The evolution of the ultrabasite massifs and their petrochemical composition has been detailed only in the publication of A.V. Razvalyaev [26], taking Sulhamid massif in the Red Sea mountains of Sudan as an example.

20

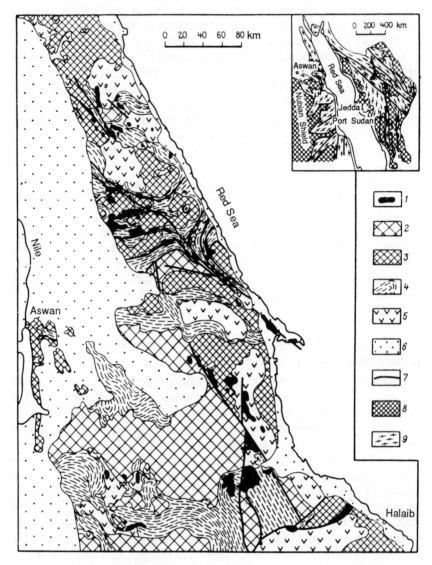

(25) Fig. 5. Tectonic plan of distribution of ultrabasites of the Arabian-Nubian shield.

1—ultrabasites; 2—'mid' massif-type continental crustal blocks consisting of upper Proterozoic-lower Palaeozoic intrusive complexes; 3—same, consisting predominantly of upper Proterozoic-lower Palaeozoic intrusive complexes; 4—upper Proterozoic green schist troughs filled with predominantly sedimentary-tuffogenic rocks (sandstones, greywacke, tuff, rarely argillaceous and arenaceous schists); 5—same, filled with basalt-andesites; 6—Phanerozoic platform cover; 7—upper Proterozoic deep-seated faults (ultrabasite furrows), activised during the Vendian-upper Paleozoic; 8—continental crust of pre-late Proterozoic age; 9—upper Proterozoic green schist volcanogenic-sedimentary folded belts.

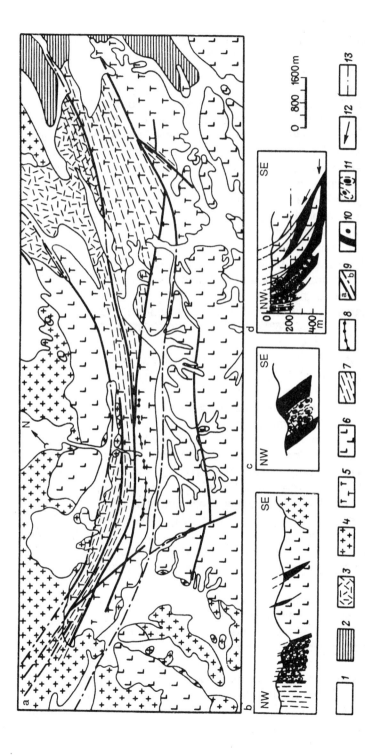

(26) Fig. 6. Geological map (a), geological section (b), constitution of mixed zone (c) and pattern of main composition (d) of the Sulhamid ultrabasite missif.

1—Quaternary deposits; 2—Neogene sandstones and conglomerates; 3—Vendian-lower Palaeozoic rhyolites; 4—late Riphean batholithic granites; 5—dunite-harzburgite complex; 6—gabbroid complex; 7—schistose highly serpentinised dunite-harzburgite; 8—listvenite bodies; 9—faults (a—proved, b—buried under Quaternary deposits); 10—tectonic plates of massive dunite-harzburgites; 11—Monomictic serpentinised mixed rocks; 12—predominant direction of tectonic movement; 13—level of contemporary erosional section.

The Sulhamid ultrabasite massif (Fig. 6) occurs in the central part of the western boundary of the Red Sea within a distance of 10—15 km from the coast. The massif extends for about 20 km and is 5—8 km wide. In plan, it has the form of an arc with a bulge towards the south-east.

(27)

In the regional plan, the massif is confined to the north-west flange of the extensive lower and middle Riphean Sudan-Arabian folded belt consisting of the thick (8—10 km) sedimentary-volcanic Nafirdeib series. Volcanic rocks, their tuffs and agglomerates of predominantly basalt-andesite composition predominate in the composition of the series. The sedimentary formation extends to the basement of the Nafirdeib series and is represented by sandstone-shale and greywacke rocks, rarely by marbles, basal and intraformational conglomerates. A general north-east strike of the folds and faults is characteristic for the belt.

The ultrabasites of the Sulhamid massif are intruded into the andesite-basalts in the form of protrusions. In the north and south-west they have pierced through the upper Riphean batholithic-granites, and in the north-east are intruded into the Vendian-lower Palaeozoic (?) rhyolites. The central and marginal parts of the massif differ considerably. The central part is represented by highly sericitised ultrabasites such as dunites, harzburgites, pyroxenites and serpentinites, spatially related to the arcuate faults separating them from the main gabbroid part of the massif. The dunite harzburgite part of the massif is highly faulted, schistose, serpentinised and split into a number of narrow linear zones reworked to various degrees. This part of the massif is characterised by the occurrence of a series of faults which conform to the strike of the massif.

Weakly reworked dunite-harzburgites have developed predominantly in the central and south-eastern parts of the massif. They represent massive dark green rocks consisting of 80—85% olivine and 15—20% enstatite and diopside. Lensoid bodies of massive chromite and asbestos occurrences are associated with them; a higher concentration of nickel sulphide (up to 0.6%) is found in the listvenitised varieties. Olivine (up to 40—50%) is present as a relict mineral in strongly serpentinised varieties. At places the olivine is so highly serpentinised that it is identifiable only with difficulty from the outline of its pseudomorphs; in such cases the rock assumes a reticulated foliate texture.

The tectonic structure of the Sulhamid ultrabasite massif is reflected in a schematic section (see Fig. 6, b). As noted earlier, intensive tectonic reworking of its central part is the main structural feature of the Sulhamid massif. This part is represented by bands or plate-like bodies of massive, weakly altered, unserpentinised dunite-harzburgite alternating with intensely reworked both serpentinised as well as tectonically fragmented mixed zones. Each such 3—20 m thick zone (see Fig. 6, c) consists of a soft, powdery, serpentinised mass with inclusions (up to 0.5 m in diam-

eter) of subangular, oval and spherical fragments and blocks. A gradual
(28) increase in 'reworking' from the massive bands to the centre of the mixed
zones is discernible. The number of clastic components concomitantly
decreases in the same direction and their form becomes more isometric. The elongation axis of finer fragments and large boudinage blocks or
'detached bodies' of massive dunite-harzburgite are oriented in a plane
with a south-easterly inclination, which is conformable with the inclination
of the massive dunite-harzburgite bodies. The inclination of these bodies
increases from the south-east to the north-west from 40°–50° to nearly
vertical in the central part of the massif. At the same time, the intensity
of tectonic deformation increases in this direction along with formation of
a thick (400–500 m) mylonite zone in the central part of the massif. The
dunite-harzburgites are intruded by a small dyke of rose coloured micro-
porphyritic granite which is conformable to the strike of the massif. The
granites are also gneissose and contain dunite-harzburgite xenoliths. All
these factors indicate that the intrusion of granites was controlled by faults
and mylonitisation of ultrabasites took place after their emplacement.

The gabbroid complex is separated by large-scale faults from the
dunite-ultrabasite zone. This complex is characterised by simpler structure
and weaker deformation. Another notable feature is that the gabbroids
contain blocks of serpentinised massive dunite-harzburgites occurring as
lensoid bodies, and striking parallel to the faults present in the central
part of the massif. The faults here have a gentler dip (about 40°) and
are thrust-like in appearance. The contact between the dunite-harzburgite
blocks and the gabbroids is tectonic in nature. Outcrops in the south-
eastern part of the massif reveal that some prefault blocks go deeper,
while others represent rootless tectonic blocks—'detached bodies' (see
Fig. 6, b).

The presence of mixed zones in the dunite-harzburgite part of the
massif implies tectonic partitioning between the massive undigested
blocks (plates), leading to the conclusion that they are tectonic in origin.
Manifold repetitions of tectonic plates and 'mixed' zones in this section
and their flattening in a south-east direction suggests the occurrence of
tectonic segments overthrust above one another.

The subvertical location of the tectonic blocks in the central part of the
massif and their flattening in a south-easterly direction became possible
under a tectonic movement in which the horizontal component of the
movement in a south-east to north-west direction predominated.

The main pattern of the composition of the Sulhamid ultrabasite massif is shown in Fig. 6, d. In the opinion of the author this pattern explains
such features of the massif as the presence of regions of tectonic sheets
and their flattening at the periphery of the massif, and also the presence
of tectonically detached bodies in the gabbroid complex.

(29) The dunite-harzburgite complex of the Sulhamid massif is similar to the continental and oceanic ultrabasites in its main petrochemical parameters; however, this complex shows maximum similarity with the oceanic harzburgites and, according to S.P. Solovev, also with the computed composition of the mantle. The main petrochemical indices of the ultrabasites—Kuno index for solidification (M) and the ratio $MgO/FeO_{average}$—vary over wide limits, indicating thereby a high degree of differentiation of the Sulhamid massif. There are varieties within these rocks with a ratio between magnesium oxide and sum of iron oxides as 4 and Kuno index as 80. This is characteristic for differentiated alpine type dunites-harzburgites and therzolites (Table 3).

In chemical composition and trend of evolution in composition, the ultrabasites of the Sulhamid massif are closer to the melanocratic members of the ophiolitic association, characteristic for the pericontinental folded structures. Relative spatial proximity with dunite-harzburgite and gabbroid complexes, presence of ferro-gabbro varieties in the latter and exclusively wide development of dunite-harzburgite components are the inherent features of the ultrabasites. Petrochemically, the gabbroids of the Sulhamid massif are characterised by a 'pure' tholeiiticline of evolution. Moreover, a small shift of the characteristic points of the gabbroid complex towards the alkali end of the diagram has been noted in the AFM triangular diagram, which may denote their formation under a relatively stable tectonic regime (Fig. 7).

Within the late Proterozoic fold regions bounded by the Red Sea, the ultrabasites can be traced along linear furrows, confined mainly to sedimentary-volcanic complexes separating the consistent, weakly altered, old Archean to late Proterozoic granite-gneiss blocks. The ultrabasite furrows are in contact with prolonged developed marginal and internal faults bordering the furrows. The predominating conformity of the faults with the boundaries of the consistent blocks leads us to suggest a palaeostructural relationship of the ultrabasites with the aforesaid furrows.

The spatial association of the majority of the ultrabasite furrows in the western structural framework of the Red Sea, containing tuffaceous sedimentary strata of lower and middle Riphean age, comprising greywackes, tuffaceous sandstones of tuffaceous shales with a subordinate amount of grit and conglomerates, draws one's attention to the course of determination of the palaeostructural conditions for the formation of the ultrabasites. The clastic material, namely, greywacke and tuff, has poor roundness, bad sorting and is represented by grains of acid effusives, tuffs, hornfels, quartz and feldspar. The conglomerates and grit have an analogous composition of clastic material. They are also characterised by poor roundness and bad sorting of grains.

Spatial association of ultrabasites with tuffaceous sedimentary strata

(30)

Table 3. Chemical composition and petrochemical parameters of the Sulhamid massif ultrabasic rocks (%).

Components	1	2	3	4	5	6	7	8	9	10	11	12	13	14	15
SiO_2	38.58	36.10	45.74	49.40	37.22	44.56	45.46	55.10	38.50	37.8	43.25	40.59	43.3	44.5	43.90
Al_2O_3	1.55	1.28	17.76	13.85	27.88	16.22	17.47	14.85	0.89	1.06	3.49	3.08	1.00	1.9	4.10
Fe_2O_3	8.50	9.32	1.91	1.36	5.01	8.14	5.67	6.66	5.04	5.79	4.74	4.65	8.4	8.7	1.10
FeO	1.29	2.51	6.75	7.25	2.66	3.41	6.97	7.97	4.38	1.01	4.37	2.49			7.90
TiO_2	0.05	0.05	0.13	0.86	0.63	1.22	2.13	1.13	0.15	0.12	0.21	0.23	0.07	0.10	0.17
MnO	0.07	0.20	0.11	0.13	0.15	0.13	0.18	0.18	0.17	0.13	0.15	0.23	0.12	0.14	0.10
CaO	0.49	3.18	13.11	9.89	15.62	8.07	9.15	8.07	0.48	0.78	3.75	2.57	0.60	1.2	3.20
MgO	36.40	32.43	10.73	11.61	5.81	9.24	5.63	2.29	43.35	38.53	36.02	34.66	45.70	42.6	37.90
Na_2O	0.09	0.33	0.72	2.26	0.22	2.40	3.88	2.70	0.25	0.12	0.30	0.19	0.10	0.20	0.52
K_2O	0.04	0.22	0.27	0.28	0.76	0.66	1.40	0.58	0.05	0.02	0.18	0.05		0.05	0.17
LOI	12.58	14.34	2.29	2.59	3.82	6.09	1.80	0.80							
Sum	99.64	100.01	99.52	99.48	99.78	100.14	99.85	100.33							
$Na_2O + K_2O$	0.13	0.55	0.99	2.54	0.98	3.06	5.28	3.28	0.30	0.14	0.48	0.24	0.1	0.25	0.69
Fe_2O_3/FeO_{gen}	0.95	0.85	0.22	0.16	0.69	0.75			0.56	0.93	0.54	0.69			0.12
MgO/SiO_2	0.94	0.89	0.23	0.23	0.15	0.20			1.12	1.01	0.83	0.85	1.05	0.95	0.86
MgO/FeO_{gen}	4.07	2.97	1.26	1.37	0.81	0.86			1.80	6.19	4.17	5.19	5.44	4.89	4.26
M	80.05	73.92	53.17	51.32	41.64	40.12	24.49	11.72	82.47	85.83	79.81	83.37	84.31	82.63	79.82

Note: Sulhamid massif: 1,2—dunites and harzburgites; 3,4,6,7,8—pyroxinites and gabbro. Average composition: continental dunites (9) and lherzolite (11), according to S.P. Solov'ev; Alpine type harzburgites (13), according to L.V. Dmitriev and others; oceanic dunites (10) and lherzolites (12), according to G.B. Udintsev and L.V. Dmitriev; harzburgite (14), according to L.V. Dmitriev and others; computed composition of the mantle (15), according to S.P. Solov'ev.

$M = MgO/MgO + FeO_{gen} - Na_2O + K_2O \times 100$ (Kuno index for hardening).

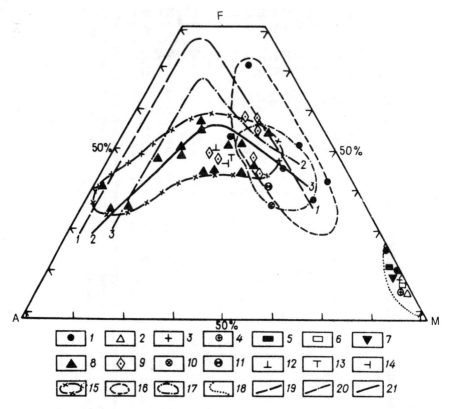

(31) Fig. 7. AFM diagram of Sulhamid massif rocks (2–7 according to G.B. Udintsev and L.V. Dmitriev).

1—ultrabasites; 2—oceanic dunites; 3—harzburgites; 4—Alpine type harzburgites; 5—continental lherzolites; 6—oceanic lherzolites; 7—computed composition of the mantle; 8—Nafirdeib andesite series; 9—Zebel-Vask basalts; 10—andesites formed in the oceanic crust; 11 same, andesitic basalts; 12—andesitic basalts formed in the sialic crust; 13—basalts of the marginal seas; 14—basalts formed in the sialic crust; 15—Nafirdeib volcanite series; 16—tholeiite basalt of the mid-Atlantic Ridge; 17—Oman diabases; 18—Oman ultrabasites. Line of differentiation: 19—Skaergard intrusion; 20—Hawaiian tholeiite; 21—Calc-alkaline series.

indicates (although indirectly) their relationship with the depressions and terrigenous facies of the latter. What was the nature of these depressions (basins) and their initial dimensions? Do they represent structural types of the internal or marginal seas or Red Sea type rifts, as held by some researchers?

(31)   The distribution of late Proterozoic depressions and their adjacent ultrabasite furrows (varied types of orientation, branching in plan) prompt the assumption that the geodynamic situation during their deposition was

characterised by multidirectional tension. The presence of terrigenous material of sialic origin in the sedimentary rocks, their shallow water origin and frequently alternating occurrence with andesites attest that, more probably, these basins represented internal or marginal seas situated in the proximity of island-arc systems having basalt-andesite volcanism. The depressions were formed on the continental crust, subjected to faulting, and then weakly altered blocks of such crusts were preserved within the folded belts (see Figs. 1, 2, 3). These troughs were subjected to compression at the transition stage of the middle Riphean to upper Riphean (32) period ($\sim$ 1000 million years), presumably because of closeness of continental blocks during the formation of fold zones and ultrabasite furrows, showing indications of squeezing and repeated flexure.

Different degrees of evolution have been observed in the ultrabasite belts depending on their position in the palaeotectonic structure. Thus, the least disturbed and least altered ultrabasite complexes formed in relatively stable tectonic conditions of the internal parts of palaeocontinents (Sudan). A lesser degree of tearing away and lesser displacement of the primary mantle substrata are characteristic of them. On the contrary, serpentinite furrows are characteristic for marginal parts of the palaeocontinent (Egypt, Saudi Arabia). The formation of such palaeocontinents took place under maximum compression and possibly they were accompanied by separation and considerable movement of the dunite-harzburgite complex, which favoured their tectonic and metasomatic transformation.

Recent publications point out the existence of important marine basins with an oceanic type crust during the late Proterozoics of the northern boundary of the Red Sea-Mozambique mobile belt and development of island arcs and ultrabasite furrows which were being successively renewed to the north-east. Such assumptions are primarily based on the assumed absence of ancient continental crust in this region.

At the present level of studies on the Precambrian of the Arabian peninsula, surmises regarding the various ages of the ultrabasite belts are disputable. The authors of the aforesaid publications also note the provisional nature and inadequate soundness of their hypothesis. But whether the ultrabasite belts of the Arabian-Nubian shield are of different ages is less a fundamental question than the nature of the basement on which the late Proterozoic furrows were formed. In other words, the main problem is whether accretion of the continental crust or its destruction took place in the late Proterozoic.

A study of the ultrabasite furrows of the Arabian-Nubian shield poses new problems arising out of the specificity of their constitution. Thus, the ultimate mechanism of movement of the mantle material to the earth's surface remains still indistinct. If it is presumed that in place of troughs having rocks of green-schist facies, there were extensive oceanic expanses or even Red Sea type rift structures, which underwent subsequent compression in the geologic past, then more extensive development of tholeiitic volcanic products and dyke complexes in folded zones could be expected. Lack of any order in the location of ultrabasite furrows, their specific tendency of occurrence and petrochemistry of orogenic subalkaline to alkaline granites, exclusively dunite-harzburgite composition of the ultrabasites, absence of any dyke complex, predominantly andesitic composition of the volcanites showing a tendency towards alkaline types and, finally, the association of the ultrabasites with graben facies prompts the suggestion that the palaeostructural conditions of formation of the ultrabasites in the northern part of the Red Sea-Mozambique belt might

(33) have differed from the typical 'spreading' model of origin in an oceanic environment, involving formation of all the members of the ophiolitic association. V. Church reached the same conclusion after studying the late Proterozoic ophiolites of Northern Africa, Arabia, Antiatlas and south-west Morocco. He also underlines that these ophiolites are confined mainly to regions with manifestation of Pan-African orogeny. N.A. Bozhko [4] and A. Kröner [42] had earlier drawn attention to this fact.

Thus, the constitutional features of the ultrabasite furrows do not exclude the possible existence of some form of stretching mechanism. In this context, the concept of shearing stress accumulation at different levels of the lithosphere, advocated by the Geological Institute of the Academy of Sciences, USSR is very interesting. The ultrabasites may have formed in relatively narrow zones and the required shear took place in the mantle with opening up of the melanocratic substratum. The proposed geodynamic environment could explain the dunite-harzburgite composition of the ultrabasites, almost total absence of deep water sediments and dyke complexes, extensive occurrence of graben and other features. The palaeobasin structures and movement mechanism of ultrabasites provide for a better understanding of the limited development of the complete ophiolitic association (pillow lava, dyke complexes, greywacke sediments) in the ultrabasite furrows which are characteristic for a typical oceanic environment. Possibly, this explains the incomplete or 'underdeveloped' ophiolitic association of the late Proterozoic rocks in the Red Sea-Mozambique belt.

The suggested mechanism of formation of ultrabasite furrows from the viewpoint of A.V. Peive's concept of stratified lithosphere is one of the possible variants. However, this does not visibly explain all the palaeodynamic environments of the ultrabasite formation in the Arabian-Nubian shield during the late Proterozoic and further does not exhaust all the modes of ultrabasite formation, particularly those which reveal no indications of considerable movement in the course of their formation. Possibly, the concept of S.N. Ivanova [10] and other geologists from the Urals on the existence of a special type of ophiolites, i.e., 'rift-generated ophiolites' are more significant for this part of the ultrabasite complexes. In their opinion, the rift-generated ophiolites formed under short periodic shear with injection of alpine-type ultrabasites under the initial stages of destruction of the continental crust, which did not lead to the appearance of oceanic structures. The formation of rift-generated ophiolites took place in narrow glacial troughs without significant separation of the continental blocks bounded by faults which penetrate into the mantle.

Many features of the Arabian-Nubian shield ultrabasites become discernible from the viewpoint of the rift-generated ophiolite concept, which could not have been considered in many cases as an ophiolitic association. This refers, first of all, to the weakly faulted massifs of the deeper parts of palaeocontinents (south-east Sudan, Ethiopia and (34) northern Kenya). A similar view exists on the nature of the lateral changes in the ultrabasites of the Arabian-Nubian shield. The appearance of a typical ophiolitic association in the ultrabasites of only northern Saudi Arabia becomes clear and attempts to identify them with ultrabasites of other parts of the Arabian-Nubian shield encounter serious difficulties. It follows therefrom that the degree of 'ophiolitic appearance' in the ultrabasites of the Arabian-Nubian shield diminishes from the north and north-east to the south and south-west, corresponding to maximum destruction and fragmentation of the marginal part of the palaeocontinent and lowering of intensity of this process in its deeper parts.

It is interesting from the viewpoint of the riftogenic nature of the ultrabasite furrows of the Arabian-Nubian shield that although they are discordant with respect to the Cenozoic rift of the Red Sea, still this territory, as it was predisposed to the destructive nature of geologic development also during the late Proterozoic, nevertheless retained this tendency still in the Cenozoic, mainly in the riftogenic stage. At this stage the Cenozoic rift-forming process did not extend over the entire region which underwent late Proterozoic destructive developments but was confined to the weakly transformed margin of the protocontinent, where maximum lithospheric inhomogeneity or maximum gradient of change in its thickness was observed. The dependence of rift formation on lithospheric inhomogeneities, which took place also in the Precambrian stage of the lithospheric evolution, might be discussed here. It is interesting to note that the Baikal rift had developed along the border of the Siberian rift, where it was preceded by destructive processes.

In view of the characteristics mentioned above, one cannot but pay attention to the fact that the late Garder rift system of Greenland (1170 million years) shows a similar tectonic set-up. It extends to the zone of branching in the Archean South Greenland craton and Ketyli mobile belt, which is subparallel to the border of the craton and is fully located within the mobile belt. Consequently, similarity of tectonic set-up of the Cenozoic rifts manifested in the border of ancient cratons and mobile belts is visibly not a random phenomenon and reflects their dependence upon lithospheric inhomogeneities which control the location of large-scale heterogeneous lithospheric blocks.

Structural Set-up of the Precambrian Arabian-Nubian Shield

Data cited in the preceding section show that the Precambrian Arabian-Nubian shield may be subdivided into complexes that can be correlated with analogous complexes in the Precambrian Mozambique belt with respect to their composition, degree of metamorphism and structural interrelationship. V.E. Khain had earlier noted a similarity between the Precambrian series of the Arabian-Nubian shield and the Mozambique belt of eastern Africa. The established nature and structural interrelationships of the gneissic and schistose complexes of the Arabian-Nubian shield help (35) in delineating the early stages of the Precambrian history of this territory. Firstly, the available data confirm its prolonged polycyclic development which was concluded during the Baikal tectonomagmatic stage. The similarity between the oldest gneiss and schistose complexes (Katarchean-lower Archean) of the Arabian-Nubian shield and the Mozambique belt reveals that they belong to the single Mozambique-Arabian belt (according to V.E. Khain) or to the Red Sea-Mozambique belt (according to E.A. Dolginov). The latter was laid in the transformed Archean (Katarchean) basement; their relicts (protrusions) are preserved in Kenya and Tanzania and have now been established by the author in the western boundary of the Red Sea and by D. Almond [35] in central Sudan.

The structural complexes of the Arabian-Nubian shield differ in unique lithology, degree of metamorphism and structural set-up. The non-uniform distribution and specificity of tectonic set-up of these complexes reflect their different structural roles in the constitution of the Arabian-Nubian shield. The lower-middle Riphean volcanogenic sedimentary metamorphic complex of green-schist facies is most widespread and constitutes fold belts of varied orientation or branches of the Red Sea belt (see Fig. 2.) The Sudanese-Arabian branch of the belt occurring in the hills of Sudan bordering the Red Sea is characterised by a predominantly north-easterly strike. It is cut off in the east by the submeridional or north-west-trending border faults of the Red Sea rift. Its north-east elongation is distinctly marked in the Arabian coast of the Red Sea. The north-easterly strike of the structure is traced up to the Rahat volcanic plateau where it gradually changes north-west, almost parallel to the Red Sea rift. The volcanogenic-sedimentary complex has a predominantly north-westerly strike in the entire north of the Arabian part of the Arabian-Nubian shield.

The eastern-Arabian branch of the Red Sea fold belt occupying the north-easterly part of the Arabian shield and consisting mainly of volcanogenic-sedimentary rocks of the late Riphean Khalaban and Murdama series has a general north-westerly strike with local deviations at a number of places, where it follows the configuration of the ancient granite-gneiss blocks of the Arabian-Nubian shield basement. The belt is complicated by a series of Naj faults striking north-west. The latter are younger than the fold structures which are characterised by significant laevo-rotation of the component of displacement.

The volcanogenic-sedimentary green-schist complex of lower to middle Riphean age in the eastern Egyptian desert constituting the northern part of the Arabian-Nubian shield shows a north-west (at places latitudinal) strike, which is different from that of the Red Sea rift. The complex constitutes troughs of varied orientation, following the configuration of old blocks of the basement and ultrabasite furrows. A submeridional strike is (36) characteristic for the volcanogenic-sedimentary green-schist complex in Eritrea and particularly in the south of Saudi Arabia. In this region they constitute a number of narrow troughs alternating with blocks of ancient granite-gneisses and occur in conjunction with the ultrabasite furrows. The Red Sea belt of the Arabian-Nubian shield split in the south and southwest into individual wedges, bordered by faults and penetrating into the ancient granite-gneiss basement of the Central African craton and the Mozambique belt.

With the establishment of the nature and role of the oldest schistose and gneissic complexes in the structure of the Arabian-Nubian shield, possibilities emerged for a full review of the problem concerning the northern boundary of the Mozambique belt and the nature of the tectono-magmatic activisation process which was extensively manifested in the African continent during 550 ± 100 million years. The tectonothermal episode of the Mozambique belt at this juncture was represented by intensive regional metamorphism, granitoid magmatism, granitisation and migmatisation. The original and inherently geosynclinal sediments and their structure had been intensively transformed as a result of such processes. This also significantly complicates deciphering the structural components of this complex and protractedly developed mobile belt. That is why the actual problem in studying the Red Sea-Mozambique belt lies in the establishment of its structural complexes and the character of their interrelationship with marginal structures and later developed rift systems.

Considerable success has been achieved in the study of the Mozambique belt owing to recent geological investigations in Kenya, Tanzania, Zimbabwe and Mozambique and also in the Arabian-Nubian shield.

The general features of structure and characteristics of manifestation of tectonomagmatic activisation have now been established within the Mozambique belt. It was found that the belt includes rocks of different 'systems' with the Dodoma series in Tanzania and the 'primitive' series in Zimbabwe being the oldest. The belt also includes charnockite gneisses of the Mozambique and Ubendi belt, where paragneisses associated with marbles and pelitic rocks are extensively developed. Mixed fold belts, submerged and partially digested by Pan-African tectonothermal episodes, are also developed in the marginal parts. The well-known conditional boundary of the western flange of the belt was drawn by J.R. Vail. However, its northern tip was inadequately studied from the viewpoint of whether data were available in literature from the time of A. Holmes for Egypt, Sudan and Saudi Arabia. At the same time, a comparison of the Precambrian of this region with the major part of the Mozambique belt is difficult due to extensive development of the sedimentary Mesozoic-Cenozoic cover (Sudan) and Cenozoic lava flows (Ethiopia).

(37) U. Kennedy introduced the term 'Pan-African tectonothermal episode' in 1964, characterising thereby tectonothermal phenomena revealed in radiological rejuvenation of ancient rocks at around 550 ± 100 million years. The contents and concept of the term have subsequently changed. This phenomenon was described by T. Clifford as 'Pan-African orogeny' (550 ± 100 million years), R. Shakleton as 'Pan-African tectonothermal episode', A. Kröner as 'Pan-African tectonic cycle' (600 ± 200 million years) and other authors in various ways. By Pan-African events T. Gass understands the cratonisation stage of north-east Africa and Arabia in the 1100–500 million years interval; he also includes the green-schist volcanogenic-sedimentary strata of the Red Sea belt here.

It became apparent in the first stage of studying regions which underwent Pan-African tectonothermal transformation (TTP*, according to N.A. Bozhko) that it is manifested in two different types of structures: 1) the cycle of sediment accumulation, magmatism, tectogenesis and regional metamorphism precedes the tectonothermal process and where the activitised process is, appears to be the logical end of this cycle; 2) it manifests only in metamorphism and intrusive magmatism. Accordingly, two types of tectonothermal transformations have been distinguished — orogenic and tectonothermal, or 'orogenically deformed upper Precambrian geosynclinal sediments' and zone of 'reactivised basement'.

According to A. Kröner, delineation of these two types of tectonothermal activisation is not fully successful, inasmuch as not only the ancient basement but also its enveloping cover were reactivated in the Mozambique belt. In addition to the two types of Pan-African episodes, he suggests segregation of a region in which only the thermal effect was responsible in accounting for the variation in isotopic ratio.

---

*Tectonothermal transformation of Pan-African—General Editor.

It is known that while initially delineating the Mozambique belt, A. Holmes had included Egypt and Sudan in it. Later, K. Schurmann acknowledged the presence of Mozambique orogeny in Egypt. J.R. Vail, D. Almond and others recently suggested inclusion of north-eastern Sudan in the Mozambique belt. Inclusion of the western boundary of the Red Sea and also of the entire Arabian-Nubian shield in the Mozambique belt is based mainly on radiological age determination data of Precambrian rocks, among which values of 600 to 480 million years are as common as the case of the Mozambique belt. However, more recent investigations have shown that geosynclinal development in this region was completed in the late Riphean-Vendian period. Highly metamorphosed Archean and early Proterozoic complexes were formed here synchronously with the Pan-African epoch and much before the advent of tectonomagmatic events. The Baikal-type role of activation was revealed in the 'rejuvenation' of these complexes and intrusive magmatism, but not in their metamorphism. The younger age of these rocks can be associated with the Baikal-type tectonomagmatic events, i.e., specific magmatic complexes correspond with the 'recorded age' of evolutionary sequences of a number of belts and appear to be characteristic links in the chain of their development.

(38) Consequently, according to U. Kennedy, applicability of younger values of radiological age determination to the boundary of the Red Sea should not be considered as evidence in favour of activation of Mozambique or Pan-African orogeny. Thus, the criteria helping A. Holmes and proponents of other fundamental concepts for including the Precambrian boundary of the Red Sea in the Mozambique belt loses significance only if it is assumed that the last result of tectonomagmatic processes took place at an interval of 550 ± 100 million years. N. Jackson emphasised this aspect with respect to the Arabian-Nubian shield and put the question: How does one treat the Pan-African events in the case that they correspond with the age of much older tectonomagmatic and sedimentary cycles, for example the Katangides, Damarides, Red Sea cycle etc., which continued active development for up to, say, 600 to 550 million years.

It is necessary to note that the criteria for distinguishing the Mozambique belt is still a matter of debate. The question is whether the intensely metamorphosed Precambrian formations are old and were later rejuvenated, or do the 'young' values of radiological age determination correspond to the existence of an independent late Proterozoic (Riphean) geosynclinal tectonomagmatic cycle. As regards the question of the northern boundary and nature of the Mozambique belt or 'Mozambique orogeny' in the boundary of the Red Sea, it is important to take into account first of all these two aspects of the problem. The metamorphic complexes detected by the author correlate with analogous Precambrian complexes of eastern Africa. The Red Sea fold belt was a component of the single Mozambique-Arabian belt during the Archean and the early Proterozoic. As for the 'younger' values of radiological age, it may once again be pointed out that they are distinctly placed within the framework of the Baikal tectonomagmatic cycle.

Given the aforesaid, the problem regarding the northern boundary of the Mozambique belt and its nature may be solved in the light of the scheme for identifying the ancient basement, which existed, as has been established recently, in the Red Sea belt and was similar to the ancient basement of eastern Africa and also through determining the correlation of the metamorphic schistose and gneissic complexes covering this basement with the distribution of the green-schist folded complex of late Proterozoic (Baikal) age. V.E. Khain had adopted such an approach to the problem much before the emergence of new data on the Precambrian of the Arabian-Nubian shield, i.e., to distinguish a singular Mozambique-Arabian geosynclinal belt. In his opinion, formation of the belt took place not later than the middle stage of the Early Proterozoic period.

Subsequently, E.A. Dolginov and others [29] developed the concept of commonality in composition and development of the Red Sea and Mozambique belts. The Red Sea-Mozambique belt is considered in their publications as comprising a heterogeneous structure and three Precambrian folded complexes of different ages have been segregated in its composition, which are separated by angular unconformities. The uneven development of these complexes reflects the lateral non-uniformity of the belt, concluding in earlier stabilisation of its southern parts and (39) subsequent rejuvenation of fold structure towards the north. From these positions, the Red Sea fold belt represents the youngest part of the single Red Sea-Mozambique belt. In essence, these publications support A. Holmes' concept on the affiliation of the meso- and catazonal altered rocks of the Mozambique belt to their ancient basement, regenerated during the period of the so-called Pan-African or Damar tectonothermal episode.

It may be noted that geologists outside the USSR have, also endorsed the concept of singular development of the Precambrian formations in the Red Sea and Mozambique belt framework. Thus, V. Pol concluded that part of the Mozambique belt is synchronous with the Red Sea belt and that the main Mozambique orogeny correlates with the Hijaz tectonic cycle (950–550 million years) of Saudi Arabia. Based on these data he has suggested that the tectonic concept of the Arabian-Nubian shield be reviewed from the standpoint of commonality of development with the Mozambique belt.

D. Almond observed a marked tendency towards broad agreement of the Pan-African episodes at an interval of 500—600 million years for the Arabian-Nubian shield and the African continent as a whole. He also observed coincidence of these episodes in the events related to metamorphism and orogenesis. This concept is also supported by I. Gass, A. Kröner, V.G. Kaz'min and others.

Following these concepts, metamorphism and intrusions in the interval of 800—450 million years are the results of radioactive heating of the continental crust, which thickened as a result of collision of lithospheric plates and closure of ocean basins. From this angle, the relevant Pan-African tectonothermal events constitute the concluding stage of development for the Arabian-Nubian shield of the Red Sea fold belt. According to D. Almond, one of the objections to such a concept of the Pan-African episode is the very large time interval for heating accompanied with thickening of the crust ($\sim$ 500 million years). On the other hand, this researcher has observed that insofar as the green-schist complexes, associated with closure of the ocean, are of extremely limited occurrence with respect to the entire Africa territory, the hypothesis linking the Pan-African events with collision of lithospheric plates can only be convincing for extreme north-east Africa and Arabia, i.e., for regions which were subjected to late Proterozoic folding. At the same time, Pan-African tectonothermal transformation in extensive regions of Mozambique, Tanzania and the Sahara remains practically unexplained from this viewpoint, which also renders this hypothesis unacceptable for the regions mentioned earlier. The concept of Pan-African events is more justified as a distinctly different phenomenon from orogenesis insofar as Pan-African heating was imposed throughout all of eastern Africa on an already existing orogenic zone, i.e., the Mozambique belt. As mentioned above, tectogenesis and magmatism in the Red Sea fold belt were synchronous with the processes of Pan-African activity and belong to their concluding stage because related to these processes by evolution and representing a natural link in the chain of events.

(40)   As for the proposed collision concept, N.A. Bozhko [4,5], having examined in detail the problem of tectonothermal transformation on the scale of Gondwana per se, has suggested that these tectonothermally transformed zones are not structures which arose as a result of collision but rather are related through evolution to the late Precambrian geosynclinal fold belts.

Accordingly, regions have been identified both in Mozambique as well as all over Africa that were subjected to development as intracratonic geosynclines during the interval of 1000—550 million years; the process was concluded by folding, magmatism and mountain building, as was the case in the Baikal fold systems (Red Sea fold belts). These regions also contain radiologically younger but earlier metamorphosed ancient Precambrian complexes, which should possibly be considered as the direct result of activisation effects of the Baikal mobile belts.

It is important to note that radiologically rejuvenated rocks in regions of the second type are superimposed and independent of the age of the substratum. Zones showing similar thermal behaviour lie much beyond the limits of the Baikal-type fold systems, and tectonic deformational processes are not characteristic of them. The entire Sudanese territory west of the Red Sea belt, including a considerable part of the Central African craton, might belong to this type of region. Possibly, they should also include the entire north African territory, occurring between the Red Sea and Dahomi-Faruzi fold belts, insofar as rejuvenation of ancient Precambrian complexes are extensively manifested here. Indirect indications for similar grouping are provided by the development of alkaline granitoid ring intrusions during the Mesozoic and the advent of volcanism in the Cenozoic, which are usually characteristic of regions showing Pan-African tectonothermal processes. Finally, activised 'rejuvenated' regions play a specific role in the Mozambique belt, where deformational features, deep-seated metamorphism, granitisation and pegmatite formation are more intensively manifested. These regions inherently occupy the central part of the Mozambique belt (Mozambique, Kenya, Tanzania). Researchers on African geology have recently emphasised the importance of distinguishing similar regions as components of large activised belts, such as Dahomi-Nigeria and Mozambique etc. In their opinion, only zones of the second and third types inherently conform to the concept of mobile activised belts. The author holds that only the first and third type zones differ fundamentally in tectonic regime. The intermediate zones represent only thermal rejuvenation of the rocks. The availability of new data will resolve the problem of whether these zones should be combined with the Baikal type or with inherently activised belts which underwent tectonothermal transformation.

In the light of the data above cited, the Mozambique belt represents a complex heterogeneous structure with prolonged polycyclic development. The Red Sea fold belt is the youngest (late Proterozoic) component of the united, i.e., Red Sea-Mozambique, mobile belt. According to (41) V.E. Khain, the development of this belt concluded synchronously with the Pan-African tectonomagmatic activisation epoch.

Thus, the Mozambique belt is subdivided, based on historical-geologic components, into inherently Baikal-type mobile belts (Red Sea fold belt) and into regions where tectonomagmatic events were synchronous with the concluding stages of formation of these belts, i.e., the Pan-African tectonomagmatic episode, which accounted for the processes of deep-seated metamorphism, granitisation and radiologic rejuvenation of ancient strata. Accordingly, the problem concerning the northern boundary of the

Mozambique belt is unambiguously solved with the acceptance of the late Proterozoic Red Sea geosynclinal fold belt as a compositional component. During the Pan-African epoch the tectonomagmatic processes within the Mozambique belt manifested themselves both in orogeny (Red Sea belt) as well as in the form of tectonothermal transformation (inherently the Mozambique belt).

It may be noted that with specific autonomy and Baikal-type structural partitioning, these belts should still be regarded as an inseparable part of the gigantic activised belts of the type Mozambique, Dahomi-Nigeria, Baikal, Grenville and others, which are characterised by a high degree of tectonic mobility and considerable heat flow during the entire history of their development. They predetermine the position of young rift structures by the nature of their disposition. The non-homogenous distribution of thermal energy within these belts is apparently related to the phenomenon that the thick geosynclinal pile of the Baikal-type area served as some sort of screen for heat energy favouring its lateral outflow, apparently along large transverse faults. As a result, the activising effect of the Baikal-type belts crossed much beyond the boundaries of these belts (for example, in the Central African craton). A similar analogy is provided by the existence of an activisation belt of the Baikal fold belt parallel to the Mozambique belt. The former is possibly concealed by the margin of the Indian Ocean. F. Dixie and subsequently V.E. Khain assumed the existence of such a fold belt. Analysis of the composition and development of the Precambrian Arabian-Nubian shield leads to the following conclusions:

1. A correlational relationship has been established between the Precambrian structural complexes of the Arabian-Nubian shield and the Mozambique belt which helps in treating the Red Sea fold belt as a compositional component of a singular Red Sea-Mozambique belt, and specifically as its youngest (late Proterozoic) branch.

2. During the Katarchean-early Archean the territory of the Arabian-Nubian shield formed the single Western Nile or Central African craton and the same underwent intense fragmentation in the late Proterozoic with the formation of diversely oriented intracratonic troughs.

3. Specificity in the development of the Arabian-Nubian shield in the (42) late Proterozoic is displayed through non-uniform destruction of the ancient craton. Its north-east part (Arabian) was subjected to maximum fragmentation where intracratonic troughs substantially transformed the ancient substratum leading to the formation of palaeobasins with an oceanic type crust. The cratonic territory occurring west of the Red Sea (Egypt, Sudan and to a lesser extent, Ethiopia) represented a relatively stable part in the late Proterozoic; the oceanic basins, on reaching closer to such cratons, gradually thinned out and finally pinched out in the form

of rift-like troughs. The north-eastern boundary of the stable part of the craton had an approximate general boundary in the form of contemporary outcrops of the oldest granite-gneiss substratum and, on the whole, roughly parallels the Red Sea. The difference in the nature of tectogenesis along this boundary allows it to be considered a furrow zone depicting the lithospheric inhomogeneities, which became imprinted even in the Precambrian period.

4. Although many models have been suggested for the geodynamic state of formation and development of late Proterozoic troughs in the Arabian-Nubian shield, the mechanisms of their formation were still not fully explained. These models range from moderate to extremely mobile variants which infer the existence of a significant number of oceanic basins in the late Proterozoic and the development of island arcs and Benioff zones in them with successive rejuvenation towards the north-east. An analysis of the Precambrian geology of the Arabian-Nubian shield shows that actual data do not always comply with these models and the role of horizontal movement was considerably magnified in them. The complete assemblage of the members of the ophiolitic association, i.e., the main indicator for the oceanic crust, is absent in the ultrabasite furrows except in the extreme north of Saudi Arabia.

5. A model based on the 'riftogenic ophiolite' concept might be an alternative for the tectogenesis of the Arabian-Nubian shield in the late Proterozoic and might have formed in a shear environment with intrusion of alpine-type ultrabasites at the initial stage of destruction of the continental crust. Riftogenic ophiolites are found to occur in the form of narrow troughs, filled in with terrigenous and siliceous sediments and bordered by faults penetrating into the mantle. In the light of this concept, many characteristics of the ultrabasites occurring south of the Arabian-Nubian shield are understandable but the same cannot be directly correlated with the ophiolitic associations.

6. A tendency towards complexity in structure has been observed in the development of the late Proterozoic mobile zones of the Arabian-Nubian shield. The ultrabasites are associated with furrow zones in the interior parts of the shield as well as throughout the African continent. Similarly, the ultrabasites are associated with shear thrusts and flexures in the boundary and peripheral parts of the shield. The late Proterozoic mobile belts demonstrate an evolutionary series of structures in which increased mobility of the earth's crust takes place from the inner parts of the African-Arabian craton to the peripheral parts with a simultaneously increasing role of horizontal movement in this direction. Correspondingly, the destruction of the continental crust was uneven: from limited shear (43) with formation of narrow, extended troughs of grooves together with the formation of 'incomplete' (riftogenic) ophiolites in the south and reaching

up to the stage of development of palaeo-oceanic environment in the north.

7. Analysis of the Precambrian Arabian-Nubian shield has helped to improve the scale and nature of the tectonothermal transformation process. It has been clarified that TTP was more widely developed in north-eastern Africa. It has been established from the late Proterozoic tectogenesis of the Arabian-Nubian shield as an example that TTP is spatially and genetically related to the formation and development of the intracratonic geosynclinal troughs, i.e., with the destruction of the ancient continental crust. Moreover, the author suggests that, at least within the Arabian-Nubian shield, the evolutionary developmental pattern of tectono-magmatic processes in the geosynclinal system should not be equated with the processes of tectonothermal and magmatic activity in the ancient basement. In other words, the TTP process in the Arabian-Nubian shield is spatially and genetically contiguous with the Baikal-type mobile belts and, apparently, was guided by the latter.

8. One of the main conclusions regarding the development of the Arabian-Nubian shield in the Precambrian vis-à-vis the problem of pre-determination of rift formation lies in the establishment of the destructive mode of its development in the late Proterozoic. Namely this territory, pre-disposed to destructive tectogenesis even in the Precambrian, retained such a tendency in the rift-forming stage as well. It is also noteworthy that Cenozoic rift formation was not localised over the entire territory of late Proterozoic destruction, but at the boundary between the stable and mobile parts of the ancient craton where maximum lithospheric inhomo-geneity has been preserved. One can perceive the predetermined forma-tion of the Red Sea rift from lithospheric inhomogeneities which emerged even in the Precambrian stage of lithospheric evolution.

2

Prerift Stage Magmatism of the Red Sea Rift Zone

(44) The characteristic feature of the Red Sea rift zone during the prerift stage of its development is extensive manifestation of alkali magmatism. Presence of alkaline intrusive complexes had priorly been established in extensive territories of eastern Africa and Arabia—from Egypt in the north to Zimbabwe in the south. The association of alkaline rocks with carbonatites played a major role in the study of the alkaline rocks of the African-Arabian rift belt. The occurrence of alkaline rocks in Africa was first noted in this belt (Malawi) and subsequently detected in a number of countries (Kenya, Tanzania, Zimbabwe and others). The interest aroused in carbonatites prompted a study of the petrography, geochemistry and structural position of alkali magmatism.

Voluminous literature dealing with the alkaline rocks and associated carbonatites of the African-Arabian rift belt, as well as covering the continent as a whole, has been published. F. Dicksie, M. Garson, B. King, D. Sutherland and L. Williams have made significant contributions to the study of Phanerozoic magmatism particularly of the alkaline type. Publications by Soviet geologists on the Phanerozoic magmatism of Africa are now available (N.V. Koronovskii, V.I. Budanov, L.S. Borodin, A.V. Razvalyaev, G.P. Shakhov, V.G. Kaz'min, V.N. Kozerenko, V.S. Lartsev and others). The Phanerozoic and particularly Cenozoic volcanism of East Africa was studied by the Soviet East Africa Expedition group, which conducted investigations within the precincts of the International 'Upper Mantle' Project.

In 1974, L.S. Borodin classified a number of provinces within the East African Belt on the basis of composition, morphology, tectonic control and spatial separation, namely, the Zambian and the Great African Rifts with eastern, western and Ethiopian rift subprovinces and the Egyptian-Sudanese province. He named the Egyptian-Sudanese province the Arabian-Nubian, because this nomenclature more accurately reflects its nature. V.I. Budanov had felt the need in 1969 for delineating an alkaline province in this region on the basis of his detailed study of the Abu-

Khuruk ring intrusion in Egypt and a number of alkaline ring complexes situated farther south. Investigations during the last decade have shown that considering the dimensions and varieties of manifestation of alkali magmatism, this vast territory needs to be delineated as an independent province.

Arabian-Nubian Alkaline Province (Fig. 8): Situated in the northern part of the African-Arabian rift belt and coinciding with the boundaries of the Arabian-Nubian shield. It includes territories in Egypt, Sudan, Saudi Arabia, North Yemen, Northern (Eritrean province) and Western Ethiopia. The geological boundaries are extensive in this framework and the eastern part of the province is particularly sharp and coincides with the Pre- (45) cambrian outcrops of the shield. The alkaline rocks of the province are extensively represented by subalkaline and alkaline granite complexes, which are known in northern and north-eastern Africa as 'young', 'rose' or 'gattar' granites. But the central type ring complexes are a more specific and characteristic form of manifestation of alkali magmatism in this province. The concept of 'Magmatogenic Ring Complex' or simply 'Ring Complex' or 'Ring Intrusives' finds place in geological literature.

Central Type Magmatic Complexes in the Arabian-Nubian Alkaline Provinces

The central type magmatic complexes occupy a specific place among platform-type magmatic complexes and represent a characteristic form of magmatism during the activisation stage of ancient platforms. The specific peculiarity of central type magmatic complexes lies in their prolonged period of formation corresponding often to one or a few large tectonomagmatic cycles (250–300 million years and more), and extremely wide spectrum of the constituting magmatic rocks, ranging from ultrabasic, basic to acidic and alkaline, which have been grouped into alkali-ultrabasic, alkaligabbroid, nepheline-alkali-syenite and alkali-granitoid formations. A combination of two to three petrographic series (often including contrasting series) reflecting the evolution of magma of different levels of generation is observed, namely, multiphase, polyformational central type magmatic complexes. Another significant feature of central type magmatic complexes is their structural-genetic relationship with the geodynamic state of stretching. The prolonged period of emergence, even within the limiting ranges of one intrusive massif, wide spectrum of petrographic associations from varying depths of magma generation and exclusive selectivity of the tectonic regime permit consideration of central type magmatic complexes as good indicators of deep-seated processes.

In 1961, Yu.M. Sheinmann identified three types of occurrence of alkali-ultrabasic magmatism: 1) marginal parts of platform; 2) zones of

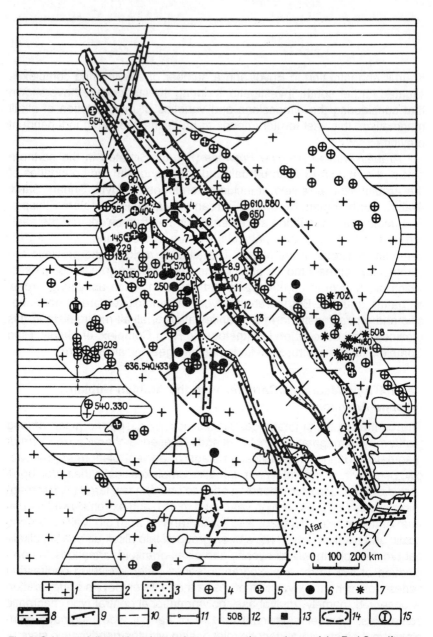

(46) Fig. 8. Scheme of disposition of central type magmatic complexes of the Red Sea rift zone.

1—Precambrian basement of the African-Arabian platform; 2—Phanerozoic platform cover; 3—Volcanogenic-sedimentary formations of the rift valley; 4–6—ring intrusives: 4—alkali-granitoid composites, 5—alkaline and nepheline syenites, 6—alkali-gabbro composites; 7—differentiated intrusives of basic composition ('layered' gabbro); 8—axial trough of the Red Sea depression; 9—rift faults; 10—transverse faults of the Red Sea rift zone; 11—faults traversed by ring intrusives; 12—radiometric age of intrusives, million years (Rb-Sr and K-Ar methods); 13—deep-water (thermal) trenches of the Red Sea: 1)—Oceanograph, 2)—Kebrit, 3)—Gypseous, 4)—Vema, 5)—Valdiviya, 6)—Nereus, 7)—Tetus, 8) and 9)—Atlantis, 10)—Sahara, 11)—Atlantis II, Discovery, 12)—Port Sudan, 13)—Suakin, 14)—range of basaltoid activisation of prerift stage of the Red Sea rift zone, 15)—main submeridional zones of deep-seated faults (I—Diib, II—Barak, III—Nile).

branching of platform and consolidated folded regions; and 3) zone of 'penetrative structural' faults. This classification of the tectonic set-up during the formation of alkaline rocks gives, like other classifications, only an idea of the spatial location of provinces. L.S. Borodin noted that it essentially encompasses all the situations of activisation of any stable region. The alkaline rocks of these regions are controlled by tension zones of the earth's crust which also include 'penetrative structural' faults. In other words, the control of alkaline petrogenesis leads to fault-forming factors. L.S. Borodin emphasises for all alkaline rocks and A.A. Frolov for (47) ultrabasic-alkaline rocks with carbonatites that, in such cases, the alkaline rocks would have been emplaced uniformly along all the 'penetrative structural' faults; however, the picture is actually far more complex.

According to the author, L.S. Borodin correctly posed the question with respect to the relationship of alkali magmatism with rifts if the rifts (excluding the Kenyan) are younger than alkali magmatism. Consequently, the prevalent correlation between tectonics and alkali magmatism cannot be explained with the help of traditional schemes and the real interrelationship might be multifaceted. It may be noted that the 'penetrative structural' fault concept is of a larger scale but nevertheless extremely indefinite, inasmuch as it unites the category of linear structures differing significantly in the nature of the crust, geodynamics and stages of tectonic development in this regard, data on new alkali magmatism provinces, particularly on the Arabian-Nubian province, are of interest because they provide scope for supplementing or correcting the prevalent concepts on the interrelationship of alkali magmatism and rift genesis.

At present, more than 100 ring intrusives have been mapped and deciphered in aerophotographs and cosmic photographs of the Arabian-Nubian province (see Figs. 8 and 9). A significant number of these intrusives are situated within the western boundary of the Red Sea (Red Sea hills of Sudan and Arabian deserts of Egypt) and were studied by geological mapping carried out by Soviet geologists, the present author inclusive. Therefore, let us first look at the intrusives of this part of the province for characterising ring intrusives (Figs. 10 and 11).

The central type magmatic complexes of the Arabian-Nubian provinces form a series by their natural association, among which three groups of complexes are differentiated: 1) gabbro-granitoid or alkaline-earth type; 2) alkali-gabbroid or alkaline type and 3) alkali-granitoid type.

Central Type Gabbro-granitoid Magmatic Complexes

This group includes complexes in which the gabbroid, and to a lesser extent, granitoid rocks predominate in the association. Alkaline rocks are usually present, although in subordinate quantity. The complexes of this

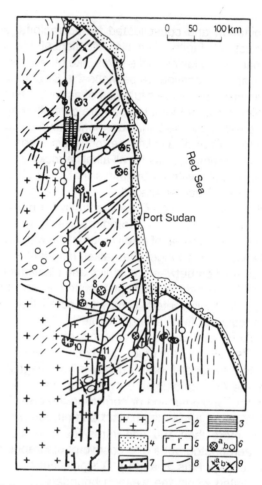

(48) Fig. 9. Scheme of disposition of ring intrusives of the Red Sea hills.

1—Archean-Proterozoic(?) schistose and gneissic complex (Kashebib series); 2—lower-middle Riphean sedimentary-volcanic complex (Nafirdeib series) and strike of the structures; 3—'Nubian sandstones' of Cretaceous age; 4—Neogene-Quaternary sediments of Red Sea valley; 5—Neogene-Quaternary basalts; 6—ring intrusives: a—investigated (1—Umm-Shibrik, 2—Ankur, 3—Salala, 4—Sasa, 5—Kur, 6—Eight, 7—Tamei, 8—Haya, 9—Takhamiyam, 10—Tekhilla, 11—Kinubanuideb)); (b—revealed primarily from deciphered data of areal and cosmic photography and individual traverses); 7—faults of the Red Sea depression and adjacent grabens; 8—regional faults; 9—fold axes (a—anticlinal, b—synclinal).

group, such as Sasa, Kur, Eight, Tekhilla, Haya and Kinubanuideb, extend to the southern part of the Sudanese Red Sea Hills.

Sasa ring complex (see Fig. 10, a): Occurs at the centre of the Sasa plain and situated about about 75 km west of the Red Sea coastal line. It

(48) has the form of an exact ring 6 km in diameter. It comprises the Archean-lower Proterozoic Kashebib rock series, represented by alternating biotitic and biotite-garnet gneisses, hornblende schists, amphibolites and marbles. A central or intraring massif and its bordering concentric ridges or ranges are distinctly manifested in the formation of the complex.

A major part of the intraring massif comprises gabbroids of olivine-gabbro to gabbro-diorite composition, showing a gradual transition to diorites towards the boundaries. Hornblendites occurring in direct contact with the carbonatites are found in the gabbro of the southern segment. Injections of predominantly leucocratic biotitic and biotite-muscovite granites are prevalent in the peripheral parts of the southern segment of the ring intrusive. The granite injection zone in the western part of the intraring massif starts with thin veins occurring parallel to the concentric ridges. Farther south, the granites are associated with biotitic diorites.

Ultrabasites, comprising intensely serpentinised dunites, pyroxenites and serpentine-carbonate rocks, are localised in the eastern and western parts of the ring intrusives. Dunites occur in the form of angular relict blocks along the ring ridges within the surrounding serpentine-carbonate rocks. The ultrabasic rocks in the western part of the massif are ring-like
(50) in form and comprise coarsely crystalline pyroxenite and hornblendite bodies 5 to 17 m wide, in contact with gabbro and carbonatites through the serpentine-carbonate variety.

The central massif occurring within the concentric ridge is represented by dioritised gabbro and uralitised pyroxenites together with albitised apogabbroic amphibolites. Systems of sublatitudinal, rarely submeridional
(51) dykes are prevalent within the ring massif and comprise mugearites, trachytes, dolerites and diabases, granites and quartz porphyry, and also stocks and annular bodies of carbonatites, pegmatites and quartz-tourmaline rocks.

The carbonatites constitute stock-like massifs of 1.6×2.4 km, closely resembling an isometric shape in plan and also open semi-circular dyke-like bodies. The stock-like massifs are emplaced within the quartz-feldspar rocks and garnet-amphibolitic schists with pegmatite intrusions. The carbonate rocks, constituting the massif, consist predominantly of white calcite which is often dolomitised. The presence of relict, island-type aggregates of short prismatic tourmaline is a characteristic feature of these carbonate rocks. The tourmaline-bearing areas are lensoid in shape and are elongated parallel to the contact. They are about 20 cm wide and a few metres long. Tourmaline forms prismatic, short tabular crystals of up to 1 cm diameter and imparts a 'speckled' appearance to these rocks.

The carbonatites of the ring dykes are represented by banded and coarsely crystalline rocks that formed at a later stage. The banded structure is caused by the dimension of calcite grains and also by the pres-

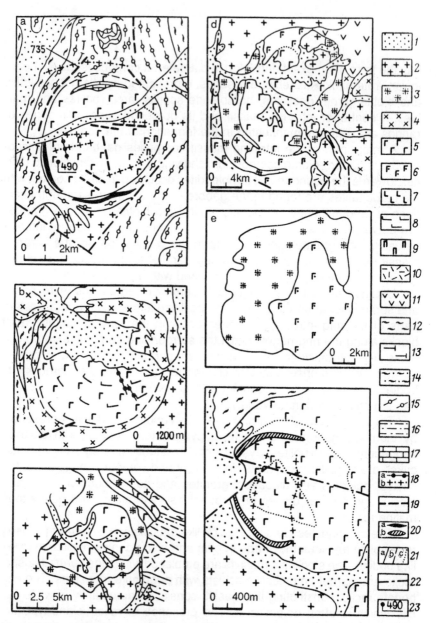

(49) Fig. 10. Central type gabbro-granitoid magmatic complexes (a—Sasa, b—Kur, c—Tekhilla, d—Takhamiyam, e—Haya, f—Kinubanuideb).

1—Quaternary deposits; 2—granites, granodiorites; 3—alkaline granites; 4—diorites; 5—gabbro, gabbro-diorites, gabbro-norites; 6—olivine gabbro; 7—magnetite-bearing gabbro; 8—anorthosites; 9—pyroxenites; 10—rhyolites; 11—17—Precambrian country rocks: 11—andesites and basalts, 12—gneisses, 13—quartz-feldspathic rocks, 14—amphibole schists, 15—garnet-amphibole schists, 16—green schist rocks, 17—carbonate rocks); 18—dykes (a—amygdaloid basalts and dolerites, b—granite-porphyry); 19—pegmatitic body; 20—dykes and stocks (a—carbonatitic, b—magnetitic); 21—geologic boundaries (a—established, b—assumed, c—facies boundary); 22—faults; 23—rock age, million years (K-Ar method).

ence of coloured bands in the fine-grained calcitic mass, giving rise to a flow structure. These banded inclusions, measuring a few millimetres to a few centimetres, consist of plagioclase (30 to 50%) and microflakes of chloritic mass (30 to 70%) with dense disseminations of opaque epidote-clinozoisite aggregates. Carbonate and quartz grains also occur sporadically in these rocks. The massive, coarse-grained carbonatites contain angular xenoliths of albitised and chloritised gabbro together with fine crystalline carbonatites of earlier generation. The main rock-forming mineral of both the textural types of carbonatites is calcite. Depending on the presence of one or the other variety of silicate, the carbonatites are distinguished as olivine- or pyroxene-bearing. The olivine-bearing carbonatite variety is highly serpentinised. The olivine has been identified as forsterite by G.P. Shakhov on the basis of its paragenetic association and shape.

Kur ring complex (see Fig. 10, b): Situated 20—25 km due west of the Red Sea coast and 8—10 km in diameter. The central massif consists of diorite along the periphery and is replaced by gabbroid and syenite-diorite varieties in the northern segment towards the centre of the massif. The massif has a concentric layered shape in the southern segment and is represented by alternate thin layers (5 to 10 m of olivine-bearing gabbro-norites, hornblende gabbro and anorthosites. The central massif is intruded by granitoids, forming dykes and veins within the massif and large ring intrusives along the periphery; the latter vary in composition from granodioritic in the south to alaskitic granites in the north. In the eastern sector of the massif, carbonatite stocks, measuring a few tens of metres and surrounded by thin pyroxenite stringers, are localised in gab-
(52) broids and syenite diorites. As in the Sasa complex, the carbonatites are represented by two textural types depicting two generations of rocks. The earlier generation carbonatites consisting of fine crystalline calcite have a banded structure resembling a flow structure. They occupy small areas among the massive, coarse crystalline replacement calcitic carbonatites of later generation.

Tekhilla ring complex (see Fig. 10, c): Situated in the vicinity of the Nile and the Red Sea basin water divide, it is in the shape of a perfect ring measuring 18 × 25 km in plan with an elongated south-west segment. The rocks constituting the complex are micaceous gneisses and migmatites of the Kashebib series (Archean-lower Proterozoic) and epidote-chlorite, hornblende schists, quartzites and marbles of the upper Proterozoic.

According to F. Delani an intraring massif consisting of olivine-gabbro has been isolated within the ring-like massif along with a concentric zone of coarse crystalline granite having an outer chilled marginal zone and the main ring intrusion. In addition to olivine-gabbro intrusives or troctolites, F. Ahmed detected 'external' olivine norite intrusives and 'internal' olivine micronorites. He also established a centripetal succession of phases in the ring granitic intrusives represented by Halokvan, Khamlal, Tainat and Tekhilla granites.

The main intrusives of granite, injected in the form of sills and conical bodies, include five varieties represented by successive phases of intrusion: 1) Halokvan granites; 2) Hamlab porphyritic granites; 3) Tainat granites; 4) Tekhilla porphyritic granites; and 5) microadamellites. The succession of injections are from the margin to the centre with the centres of the intrusions shifted south-west. The granite intrusives have vertical contacts which are well exposed in the northern and southern part of the concentric ridge. The granites are very similar in their petrographic and chemical composition. They consist mainly of feldspar (65–69.5%), quartz (24.5–29.5%) and a subordinate amount of biotite and hastingsite; the texture is massive, porphyritic or gneissic and is due to preferential orientation of minerals. The microadamellites form discontinuous narrow bodies with about 70 m maximum thickness and branch out into smaller bodies dipping to the centre of the complex at 40° to 65°. The rock is fine-grained and often has porphyritic and flow structures.

The petrochemical composition of the Tekhilla complex varies from weakly alkaline to alkaline. The granites are moderately alkaline. Increase in alkalinity has been observed in granites with the intrusion of younger phases. The formation of the Tekhilla magmatic complex, according to the K-Ar age determination method, encompasses a 440–550 million year interval.

Tahamiyam and Haya ring complexes (see Fig. 10, d, e): Respectively 18 and 11 km in diameter, their central masses consist of olivine gabbro. The Tahamiyam concentric and the Haya semi-circular intrusions (53) are subalkaline granites. The latter especially displays a predominant development of semi-circular granites from the enclosing granitoid rocks and absence of granites from the associated diorites.

Kinubanuideb ring structure (Fig. 10, f): The smallest among the concentric intrusions (diameter 1.5 km), it is situated on the left bank of the tributary of the same name flowing from Vadi Langeb and is distinguished by simple composition. It is confined to small intrusions of Palaeozoic (?) gabbroids injected into the Archean-lower Proterozoic gneissic complex along the intersection of submeridional and north-east-trending faults. A fine and moderately coarse crystalline gabbro massif, measuring 1.5 km in diameter, is eccentrically emplaced within the coarsely crystalline gabbro intrusive of isometric shape in plan (diameter 3 km). Gabbro is saturated with magnetite in the central part of this massif, and shows sideronitic texture at some places. A conical 0.6 to 3 m thick magnetite dyke can be traced along two-thirds of the perimeter of the small massif periphery. The core surface dips to the centre of the massif at 45°.

Central Type Magmatic Complexes of Alkaline Composition

The syenite group of rocks, namely, trachyte, alkali-gabbroids or sub-alkaline granites, belonging to the meso-abyssal or subvolcanic phases of depth according to A.A. Frolov's classification, predominate in the association of the described rock complexes. These complexes are predominantly developed in the north of the alkaline rock province, although singular alkaline and granitoid ring intrusion (Tamei, Odrus) occur in areas with predominant gabbro-granitoid complexes. In all, 13 ring complexes of such types have been identified. Among them, eight are situated in the northern Egyptian part of the Red Sea hills and have been described by El Ramli, V.I. Budanov and N.E. Derenyuk. These are intrusions occurring in Abu Khuruk, El Kahfa, El Naga, Nugrus-El Fokani, Mishbekh, Mansuri, Tarbtie and El Gezira [34,40]. They vary in diameter from 4 to 9 km. The common feature for this group is the presence of relict volcanic cones composed of trachytes, trachy-basalts, phonolites, latites, rhyolites and pyroclastics of the same composition which help in their grouping under subvolcanic facies of depth. The enclosing rocks are either metasedimentary-volcanic formations and batholithic granites and diorites of upper Proterozoic or Katarchean-lower Archean gneisses.

The presence of hypabyssal alkaline varieties is characteristic for the complexes occurring in metasedimentary and volcanogenic rocks (El Gezira, Mansuri and Tarbtie). Excluding the El Kahfa complex whose central stock is composed of essexite gabbro, the central massifs of other complexes of this group consist of hypabyssal varieties of alkali syenites (El Gezira), stocks and carbonatite dykes (Mansuri) or a system of (54) concentric alkali-syenite dykes. The ring intrusives of these complexes comprise alkali syenites (El Kahfa and El Gezira) or their hypabyssal analogues (Tarbtie).

The central stocks in ring complexes occurring among gneisses and syntectonic granites, granodiorites and diorites (Abu Khuruk, El Naga, Nugrus-El Fokani and Mishbekh) comprise syenites and alkali gabbroids (El Kahfa) or leucocratic essexite (Nugrus-El Fokani). The external ring intrusive complexes of Abu Khuruk, El Kahfa and Mishbekh also comprise alkaline syenites, the Nugrus-El Fokani complex comprises alkaline granites and the El Naga complex consists of phenitised porphyroblastic gneisses. Solvsbergite-tinguaite and bostonite-grorudite rock series together with chydolerites and phonolites have been noticed in dyke rocks.

Four ring complexes of the group discussed above are situated in the Red Sea hills of Sudan (Salala, Ankur, Kelly and Tamei). Notwithstanding the predominance of gabbroids in the rock association, the imperfectly developed concentric complex 'Eight' may be included within this group, insofar as it belongs to the subvolcanic facies in respect of content of hypabyssal and effusive rocks.

Salala ring complex (see Fig. 11, a): Occurring 120 km due west of the Red Sea in the northern part of the Red Sea hills among the upper Proterozoic volcanic Nafirdeib rock series which are intruded by batholithic diorites of the upper Riphean age. The complex is bifocal in composition and consists of large and small concentric complexes.

The large or main complex is constituted by a narrow, open outer ring intrusive, an inner ring intrusive and a central stock. The outer ring intrusive has a near oval shape in plan with axes measuring 7 and 8.5 km; the thickness is 0.5 to 2.4 km (in the west); the contacts dip steeply from the centre of the complex. The intrusive comprises leucocratic, essentially soda-aegirine syenites. The internal ring intrusive of anorthosites constitutes a major part of the concentric massif. The anorthosites are massive, coarse-grained and consist mainly of andesite, labradorite and biotite; a high concentration of prismatic apatite (up to 10%) is characteristic of them. Rocks of the internal ring are intersected by dykes of smaller dimensions and consist of alkali-feldspar-bearing granites and solvsbergites.

The central stock of the main complex is represented by separate outcrops comprising troctolite in the northern part and later (250 million years, K-Ar method) formed nepheline syenites in the south. The nepheline syenites contain large xenoliths of anorthosite and hornblendised dolerites which imparted considerable inhomogeneity to their composition. The rocks are leucocratic if xenoliths are absent and contain large (up to 1 cm) nepheline grains, which sometimes account for 30% of the composition. The dyke series of the main complex is represented by trachytes, alkali-feldspar-bearing granites, solvsbergites, grorudites and dolerites.

(55)   The small ring complex has been intruded in the south-west segment of the main complex and has a similar weakly elongated oval shape in the north-east direction with axes measuring 3.6 and 4.8 km. Like the main complex, the small complex consists of a narrow external and a wide internal concentric intrusive. The former is open in the north and at this place the ring is marked by a sickle-shaped fault. The external concentric intrusive is composed of rose-coloured syenite with a considerable amount of (10 to 15%) hematite, often forming schlieren bonds and veins at places. The syenites contain inclusions of hornblendised xenoliths from enclosing rocks together with their intersecting dykes. They consist mainly of perthitic and antiperthitic orthoclase, and, to a lesser extent, of microcline. The internal ring-like intrusion of the small complex is inhomogeneous in composition. The chemical composition of its constituent rocks is that of the group of melanocratic alkaline rocks, corresponding to alkali gabbro (Figs. 12, 13).

The conical central stock is oval-shaped in plan, measuring 1.5 × 2 km, and elongated in a north-easterly direction. The stock is situated

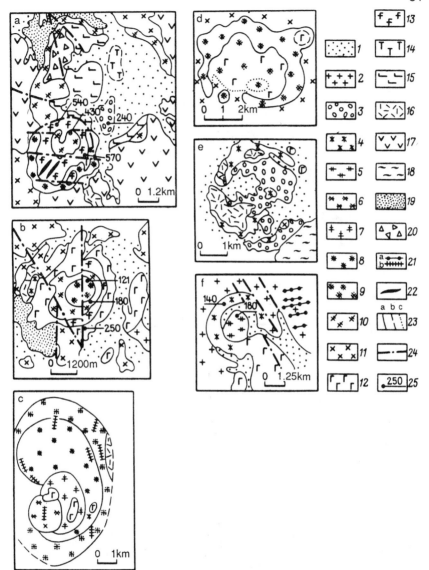

(50) Fig. 11. Central type magmatic complexes of alkaline composition (a—Salala, b—Ankur, c—Kelly, d—Tamei, e—Abu-Khuruk, f—Umm-Shibrik).

1—Quaternary deposits; 2—granites, granodiorites; 3–10—syenites: 3—nephelines, 4—alkali, 5—quartz, 6—riebeckites, 7—hastingsite, 8—fayalites, 9—hematitite, 10—aegirine; 11—diorites; 12—gabbro, gabbro-norites, gabbro-diorites; 13—alkaline gabbro; 14—troctolites; 15—anorthosites; 16—rhyolites; 17 to 20—Precambrian country rocks: 17—andesites and basalts, 18—gneisses, 19—green-schist rocks, 20—agglomerates; 21—dykes (a—amygdaloid basalts, dolerites and bostonites, b—syenites); 22—carbonatite dykes; 23—geologic boundaries; 24—faults; 25—rock age, million years (K-Ar method).

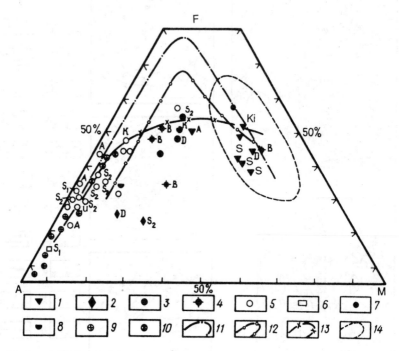

(55) Fig. 12. AFM diagram of ring intrusives in the Red Sea hills of Sudan.

1—gabbro and gabbro-norites; 2—anorthosites; 3—alkali-gabbro; 4—diabase and dolerite dykes; 5—syenites; 6—nepheline syenites; 7—basalts (180 million years); 8—trachytes of the Natash series (chalk); 9 and 10—trachytes and bostonites of Red Sea hills: 9—Mesozoic (85 million years), 10—Carboniferous-Permian (290 million years). Line of differentiation: 11—Skeargard intrusion; 12—Hawaiian tholeiites; 13—Red Sea ring complexes; 14—basalt fields in mid-oceanic ridges. Intrusives: (S_1—Salala (main complex), S_2—Salala (minor complex), S—Sasa, K—Kelly, Ki—Kur, U—Umm-Shibrik, A—Ankur, B—dyke complex, D—Dally composition).

(56) eccentrically in the northern segment of the ring and comprises red, coarse-grained syenite, consisting of albite (70%) with chessboard twinning replacing orthoclase, and is significantly poorer in silica than all the previously described syenites.

The vein rock series of the smaller ring complex is represented by quartz-free pegmatites, leucocratic syenites and carbonatites. Besides the veins, the carbonatites also constitute two stock-like bodies intruding into the intraring massif of the south-east sector. They have a circular (10 m in diameter) and oval (50 to 30 m) shape in plan and comprise serpentine-carbonate rocks composed of calcite and dolomite crystals. G.P. Shakhov distinguished magnetitic and apatite-magnetitic carbonatites depending on the proportion of disseminated minerals. Calcite-magnetite (in the

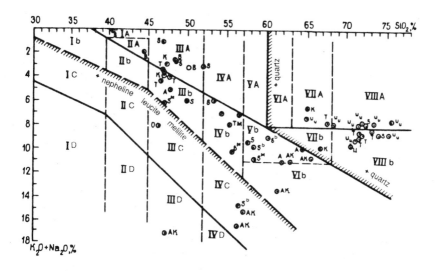

5) Fig. 13. Distribution of composition of rocks of the Red Sea ring magmatic complexes in the petrochemical groups of magmatic rocks (according to A.A. Marakushev).

Rock groups: IA—dunites, peridotites; IIA—picrites; IIIA—gabbro, alkaline pyroxenites; IVA—gabbro-diorites and pyroxenites; VA—diorites; VIA—quartzose diorites; VIIA—granodiorites; VIIIA—granites; IB—kimberlites; IIB—alkaline picrites; IIIB—alkali-gabbroids, IVB—monzonites and trachy-andesite basalts, VB—syenites and trachytes; VIB—alkali syenites; VIIB—quartzose syenites; VIIIB—alkaline granites; IC—jacupirangite; IIC—basanites; IIIC—theralites, tephrites; IVC—nepheline, syenites, phonolites; ID—turjaite; IID—ijolites; IIID—nephelinites; IVD—khibinites. Intrusives: A—Ankur, AK—Abu-Khuruk, B—Baraka, K—Kur, O—Odrus, S—Sasa, S^b—Salala (main complex), S^m—Salala (minor complex), T—Tekhila, TM—Tamei, U—Umm-Shibrik, U_u—Ubeidjah.

west) and quartz-hematite (in the east) ore bodies occur in the exocontact regions of syenite ring intrusives of the smaller concentric complexes.

Ankur ring complex (Fig. 11, b): Situated on the western slope of Vadi-Diib. In plan, it is elliptical in shape (3 × 4 km) and elongated in the meridional direction. Upper Riphean diorites constitute the major part of the complex, and the metasedimentary Nafirdeib rock series (lower-middle Riphean) occur only in the south-west segment.

The complex is composed of external and internal ring intrusives and
(57) central stock. The exocontact of the external ring intrusives is almost vertical and gently dips from the centre while the internal intrusive dips towards the centre of the complex. The latter is composed of gabbroids, varying in composition from troctolites to microgabbro. Gabbro occurs to the eastern part of the external ring. Microgabbro occurs between the external and internal rings in the form of a semi-circular massif in the eastern part of the complex and represents a facies variant of gabbroids of

the external intrusive. The internal ring intrusive has sharp contacts which are often tectonic in nature. Its external contact dips steeply towards the centre of the complex while the internal contact is vertical. The intrusive is composed of aegirine syenites which enclose large xenoliths of altered gabbro. The central stock is composed of quartzose syenites with medium to coarsely crystalline, xenomorphic to hypidiomorphic granular texture.

Emplacement of the Ankur ring intrusive complex took place from the periphery to the centre. The age of the external gabbroid ring is 250 million years (K-Ar method) and that of the central stock 121 million years. This means, following A.A. Polkanov's scheme, the Ankur ring complex belongs to the central type.

Kelly ring complex (Fig. 11, c): Located within the Butan green schist rock series, it is elliptical in shape with axes of 1.6 km and 3.3 km and consists of three ring intrusives with an eccentrically situated central stock. Gabbro is the earliest differentiation product of the intrusive and is represented apparently by large xenoliths of the first phase of the complex. Next in order of formation is the external ring intrusive (chilled zone according to F. Delani) composed of quartzose, augitic syenite and porphyritic microgranites. These formations are in contact with the rhyolites from the external side. The latter are considered extrusives, whereas the gradual transition to syenites, described by F. Delani, indicates that the rhyolites belong most probably to the chilled zone of syenites rather than the extrusives. The middle ring is composed of fayalite-bearing syenites, the internal ring contains hastingsite-bearing syenite and the central stock comprises riebeckite-quartz syenites, at places with pegmatoid structure. Syenites in all the rings are intersected by radial and arcuate dykes of microsyenites, bostonites, grorudites and aegirine-felsites.

Tamei ring complex (Fig. 11, d): Situated on the right bank of Vadi Odrus, flowing parallel to the meridional Diib system of faults. The complex intrudes into the upper Palaeozoic diorites; it is oval in shape, measures 4.5 × 6.0 km and is elongated sublatitudinally. Its outer ring intrusive body is composed of leucocratic syenite-porphyries, whereas the central massif consists of monzonite syenite. A region (1.5 × 1 km) of coarsely crystalline magnetitic syenites, containing up to 20% magnetite, can be identified within the syenites.

Abu-Khuruk ring complex (Fig. 11, e): Investigated by El'Ramli, V.I. Budanov, N.E. Derenyuk, A. Gindi and others, it has been cited as an example of alkaline ring intrusive in the northern part of the
(58) Arabian-Nubian province where relicts of volcanic formations have been preserved. It is 4 km in diameter and constitutes concentric syenite intrusives with nepheline syenite internal massif. Syenites vary from grano-syenites (external part of intrusion) to alkali syenites. Conical bodies of nepheline syenites with umptectite as one of their variety and the

central stock of diorites together with urtite at some places, have been differentiated in the internal massif. The presence of alkaline soda-rich trachyte is characteristic of the volcanogenic rocks of essentially rhyolitic composition.

Umm-Shibrik ring complex (Fig. 11, b): Situated 30 km due north of the Ankur ring complex, it intrudes into granites and gabbro of the upper Proterozoic, is 3.6 km in diameter, composed of central stocks of grano-syenites, and 180 million years old (K-Ar method), but together with the ring intrusives 140 million years of age. The syenites are microscopically analogous to the syenites of the Ankur complex. The presence of grano-syenite stocks in the complex suggests that it is transitional to the group of alkali-granitoid ring complexes.

Central Type Alkali-granitoid Magmatic Complexes

This group is composed of syenites and granites, and petrochemically extreme members of this series are the alkali granites only. The group is represented by Odrus and Umm-Shibrik intrusives within the Red Sea boundary, and beyond this zone by the Sabaloka and Bayuda intrusives in the Nile River valley; Bakhari and Uveinat intrusives in Libya; Ueidjah, Khadb Aldiakhin, Midian and others in Saudi Arabia.

Ueidjah ring complex: Situated in the eastern part of the Arabian shield and studied in detail by F. Dodge. It may illustrate the composition and development of analogous alkali-granitoid groups of intrusives developed along the periphery of this shield. The concentric intrusive is composed of Al Arief granodiorites, varying from quartzose monzo-diorites to granites in composition. The granodiorites are intruded in the central part by Taaban granites and by the concentric Abu-Khurg 'leucogranite' dyke, measuring 15 to 20 km in diameter and up to 2 km in width. This dyke has both an intersecting as well as conformable injection-type (layered) contact with crystalline schists. The enclosing quartzose hornfels and biotite schists are vertical and occur conformably with the intrusive.

The intrusives Sabaloka in the Nile River valley and Bakhari in the Uveinat Massif (Libya) are simpler in composition. They are represented by ring dykes of granophyre and granitic bodies respectively.

'Layered' gabbro massifs: Ultramafic and mafic rocks, extensively developed within the Red Sea and Aden Bay framework, are described under this heading. Intrusive massifs of this group are represented by (59) peridotites, pyroxenites, gabbro, norites and anorthosites belonging to the peridotite-pyroxenite-norite formation or, according to Yu.A. Kuznetsov, to the differentiated gabbro and noritic intrusives. At the same time, the layered gabbroid intrusives display petrochemical and mineralogic similarity with gabbroid intrusives of concentric magmatic complexes. Considering

the wide variations of the gabbroid intrusive rocks of the group under discussion and their inadequate study, particularly in Arabia, their formational grouping needs to be confirmed. However, this situation is not significant for the present analysis and these complexes may be combined into the 'layered' gabbro group—a term by which they are known in literature published outside the USSR on the Arabian-Nubian shield.

The massifs of basic rocks have the shape of lopoliths and show a distinct concentrically zoned structure. Titano-magnetite, ilmenite-magnetite and at places copper-nickel sulphide ore associations are characteristic features of this massif.

The 'layered' gabbros in the Arabian-Nubian shield are localised within the Red Sea-Aden Bay framework. They are more extensively developed in Saudi Arabia, Egypt, Sudan and on both sides of Aden Bay in the People's Democratic Republic of Yemen and northern Somali. In Saudi Arabia, the 'layered' gabbro massifs are oval in shape with a diameter ranging from 0.8×3 km to 2.9×8 km and, at places, 10×24 km. In most cases, they pierce through the green-schist strata of lower-middle Riphean age, forming synclinorial structures within the gneissic blocks of the basement or large granite batholithic massifs. They are composed mainly of clino-pyroxene containing gabbro with or without olivine, norites, troctolites and anorthosites. Cumulative and intercumulative constituents are well exhibited in these rocks. Plagioclase is the predominant cumulative mineral—forming norite with orthopyroxene and troctolite with olivine.

In the African part of the Arabian-Nubian shield, the 'layered' gabbro massifs are developed in the Arabian desert of Egypt and in the Red Sea hills of Sudan.

The Abu-Galaga and Akarem 'layered' gabbro massifs have been studied in greater detail in the Arabian desert of Egypt. Ilmenite-magnetite deposits are associated with the former and copper-nickel-cobalt mineralisation is related to the latter. The massifs are lopolithic in shape. The Abu-Galaga massif is composed mainly of gabbro-norite and hypersthene-gabbro, with a subordinate amount of melanorites and anorthosites. The massif formed in three successive phases: magmatic, metamorphic or mobilised and hydrothermal. The last is associated with the final phase of ilmenite ore genesis. The composition of the Akarem massif is similar to that of the Abu-Galaga massif. The former differs from the latter by the presence of peridotites constituting about 30% of the massif. Interlayering of peridotites and pyroxenites with gabbroids has been noticed in sections. Copper and nickel sulphides occur as disseminations in all the basic and ultrabasic rock types. Two intrusive phases have been identified in the course of their formation. The first is the formation of gabbronorite lopolith with separation of more melanocratic varieties. Injection of peridotite, plagioclase peridotite, melanorite and pyroxenite dykes with

(60)

complex copper-nickel-molybdenum mineralisation is associated with the second phase.

In the Red Sea hills of Sudan, basic rocks participate in the formation of multiphase, often polyformational ring intrusives. A number of concentric intrusives are distinguished here, the central parts being composed of gabbro, norite and anorthosite. These are Kur, Haya, Takhamyam, Tekhilla and Kinubanuideb and other intrusives (see Fig. 13). The Kur and Kinubanuideb intrusives show a more distinctly developed concentric layered composition; thin (5—10 m) alternations of olivine gabbro-norites, hornblendic poikilitic gabbro and magnetite bodies are found in them.

Radiometric age determination (K-Ar and Rb-Sr methods) shows that the 'layered' gabbro massifs of the Arabian-Nubian shield are mainly of 702—415 million years of age. R. Coleman, R. Flesk and other researchers estimated 20—23, 769 and in a singular case, 1374 million years of age for individual massifs. These massifs were thus formed in the late Riphean, Vendian and early Palaeozoic periods while individual massifs (At-Tarf) confined to the eastern marginal fault of the Red Sea rift formed in the Miocene. These data show that formation of the 'layered' gabbro intrusives in regions adjacent to the Red Sea took place over a long period of time. The age of 1374 million years seems to be the only age value whose real significance is not clear. It cannot be excluded that it corresponds to the beginning of formation of 'layered' gabbro in regions manifesting early stabilisation of structures of the Red Sea folded region, as was suggested earlier. Approximately such an age range (truly, without the Palaeozoic alternative) has been established for the 'layered' gabbro of the Ubendi zone in north-west Tanzania.

The general pattern of distribution of 'layered' gabbro intrusives and also conditions of localisation of individual massifs assume great interest. They display a distinct structural control that exhibits a relationship with both the ancient folded complexes and the much later tectonic forms. The 'layered' gabbro intrusives of the Arabian-Nubian shield and also the concentric massifs are confined to the network of faults striking north-east, north-west, submeridional and, to a lesser extent sublatitudinal as well as the points of intersection of these faults. However, the coincidence of their occurrence with the Cenozoic rift systems of the Red Sea, Aden Bay and Rukva in Tanganyika deserves particular attention. The remarkable uniqueness of ancient endogenetic regimes and magmatism along definite directions is thereby revealed, wherein large faults subsequently became the zones of rift formation.

Gattar subalkaline and alkaline granitic complexes: Extensively developed in Egypt, Sudan, Saudi Arabia and, to a lesser extent, in Ethiopia, (61) People's Democratic Republic of Yemen and Somali. Greater interest in these granites is explained by the fact that industrially significant deposits of tantalum, niobium, tin, wolfram, beryl, rare earths and other metallic minerals are associated with them. Subalkaline and alkaline composition is a common feature for all the granitic intrusions. The age range of their formation is wide. Thus, they include 120—170 million-year-old granites in Nigeria and 295 million-year-old granites in Algeria (Air). Their age ranges from 620—450 million years in Egypt, Sudan and Saudi Arabia. Isolated intrusives in Sudan, Ethiopia and Somali show a younger age ranging up to 50 million years.

Within the Arabian-Nubian shield, the gattar granites were more thoroughly studied in Egypt and Sudan. At these places they are associated with rare earth and tin-wolfram mineralisation. With respect to the Vendian molasse formation, the Khammamat gattar granites have been subdivided into two age groups: lower—pre-Khammamat or early gattar and upper—post-Khammamat or late gattar.

Systematic investigations of the eastern Egyptian deserts, participated in by Soviet geologists, helped to confirm the sequence of formation of the 'young' granitic complex, to supplement significantly their petrographic characteristics and to establish their tectonic position. It was shown that granites of both age groups (complexes), particularly the younger one, have inherited superimposed processes of albitisation, feldspathisation, greisenisation, silicification and associated rare metal mineralisation.

Granites of the early gattarian complex are represented by medium-grained biotitic and biotite-hornblende, granophyric varieties, granite-porphyries and felsite porphyries. They are composed of relatively large massifs occurring discordantly with fold structures in the central part of the eastern desert. Granophyric granites occur in close association with the medium-grained biotitic granites. Felsite-porphyries are found in the endocontact parts of the massifs. Medium-grained biotitic granites are associated with dykes of fine-grained biotitic granites, granite-porphyries and felsite porphyries.

The important types of the late gattarian complex are coarse-grained biotitic and fine-grained leucocratic granites. Muscovitisation and albitisation are particularly well displayed in them. Coarse-grained biotitic granites are rose or reddish-rose fully crystalline rocks. The characteristic morphological feature of these granitic massifs are their amenability to form sharp peaks in relief, owing to which they are easily diagnosed in any locality and are deciphered well in aerophotographs. They consist of quartz (25 to 30%), microcline-perthite (50 to 60%), albite-oligoclase (10 to 20%), biotite (5 to 10%) and hornblende (less than 3%). The accessory minerals are zircon, apatite, tantaloniobates, magnetite and ilmenite. The sharp predominance of potash feldspar above the plagioclase distinguishes them from early gattar granites. Large granite massifs have a zonal structure

(62) with separation of coarse-grained biotitic granites in the central zones and biotite-muscovite and alaskitic granites showing extensive development of albitisation in the outer exocontact zones.

The fine-grained leucocratic granites are spatially associated with coarse-grained biotitic granites. These leucocratic granites usually form steeply dipping (50° to 80°) dykes and sills extending from a few tens of metres to a few kilometres. Albitisation and muscovitisation processes are also characteristic for them.

Muscovitic granites and anorthosites occur in close spatial association with the late gattar granites and show tantalum-niobium, beryl and tin-wolfram mineralisation of industrial significance.

Muscovitic granites comprise small stocks, sills or layered bodies. They usually comprise marginal parts of coarse-grained biotitic granite massifs, of which they are the metasomatic products. A gradual transition from biotitic granites to muscovitic granites has been observed in a number of massifs. In such cases the texture and structure of the granites and preserved, only their colour changes, imparting a leucocratic tone to the muscovitic granites.

The apogranites represent more metasomatically altered late gattarian granites. The metasomatic processes, related to postmagmatic solutions, are represented in the form of intensive albitisation and muscovitisation. Depending on the dimensions, shapes, erosional depths and metasomatic zoning, different degrees of ore content are observed in the apogranites of the eastern Egyptian desert. V.A. Bugrov established that rare metal content increases in apogranites with an increase in the dimensions of massifs, decrease in erosional depth and flattening out of their contacts with enclosing rocks.

The following pattern has been established in the disposition of the gattar granitic intrusive complex in Egypt or the 'younger' granitic complex within the Arabian-Nubian shield: a major part of the intrusives is confined to regions with a folded late Proterozoic green-schist complex and, to a lesser extent, to its Archean-lower Proterozoic granite-gneiss basement. Emplacement of granites is controlled by the folded structures and faults of north-west and north-east and, to a lesser extent, sublatitudinal orientation which border the interblock mobile zones and their feathering joints. The early gattar granites are localised predominantly in folded blocks which were affected by uplifting movement during the geosynclinal stage. The late gattar coarse-grained biotitic granites are confined to activised faults of interblock volcanogenic-sedimentary Riphean troughs or to the marginal parts of the blocks. They are not characteristic for the central parts of these blocks. Distribution of muscovitic and leucocratic

(63) granites and felsite porphyries are controlled by cross faults of much later origin, which are discordant to the blocks and separated from them by fur-

row zones. The cross faults have a predominantly north-west and rarely north-east strike.

The extensively developed 'red' and 'rose' alkaline and ultra-alkaline granites of the Arabian Peninsula in Saudi Arabia and Yemen, comprising mainly the concentric intrusives, stocks and dykes of 620–520 million years age range belong to these 'young' granites.

The subalkaline and alkaline granite intrusives are less widespread in further southern parts of the Arabian-Nubian shield within Ethiopia and adjoining the Precambrian basement (Yemen, northern Somali) of Aden Bay. Besides the decrease in intensity of alkali-granitoid magmatism in the southern direction, this aspect may be explained by the fact that these areas are so far inadequately studied and characterised by extensive Cenozoic lava flows. The intrusives are represented in Ethiopia by numerous small bodies of microgranites, alkaline granites and granite porphyries. According to V.G. Kaz'min, the granites occurring along the western boundary of the Ethiopian plateau are of late Palaeozoic age. They are represented by coarse-grained porphyritic biotite and biotite-hornblende-bearing massive varieties. Morphologically, the granites in the central and northern flanks of the Arabian-Nubian shield are, to a great extent, represented by isometric bodies with circular outline in plan and are controlled by faults. The alkaline granites of the gattarian complex in the Arabian-Nubian shield constitute complex multiphase concentric intrusives of alkaline and alkali-gabbro composition together with the homogeneous massifs.

Until recently, information on the petrochemistry of the gattar granites was extremely meagre.

A petrochemical analysis of the subalkaline and alkaline granites was conducted by the author for virtually the first time in north-eastern Africa, based on recalculation of more than 150 chemical analyses of granitoids in Egypt and Sudan. Part of these analyses were taken from literature, and the rest are the author's original data. These data enabled designation of parental magma and its evolutionary path for the more thoroughly studied granitoids of Egypt and north-east Sudan. For a more specific determination of the petrochemical affiliation of the specific granite types of north-eastern Africa, they have been compared with 'young' Nigerian and Algerian (Khoggar) granites, and also with the standard geochemical types of granitoids in Trans-Baikal and the Mongolian People's Republic (Table 4).

Recalculation of chemical analyses revealed the following petrochemical features of the gattar granitic complexes. The Kuno diagram, illustrating the ratio between sum of alkalis and silica, shows their affiliation to subalkaline high alumina and high alkaline series (Fig. 14). Based on the distribution of compositional points in the Wager diagram, it was found that

(64) the course of crystallisation differentiation extended from batholithic granites to early and late gattar granites (Fig. 15). The salic components which shifted the evolutionary trend of the magma towards increased alkalinity had a leading role in its differentiation. During the course of its differentiation, the granitoid magma became impoverished in iron and magnesium and the alkaline content increased in the terminal stages of fractionation. Starting with the early gattar granites, potassium predominates among the alkaline components and this difference increases towards the final stage of differentiation. The sum total of alkalis exceed 8% in early gattar granites with an approximately equal ratio of K_2O and Na_2O. The late gattar granites are potassic (K_2O + Na_2O = 1.10 – 1.15), the sum of alkalis attaining 9 to 10%. The coefficient of agpaiticity also displays the same tendency.

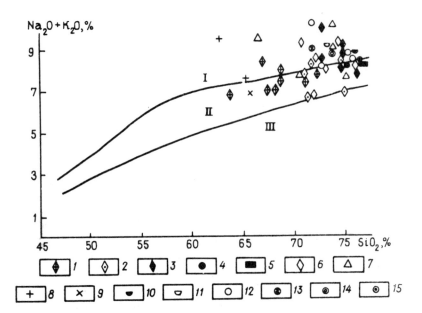

(65) Fig. 14. Correlation diagram for Na_2O + K_2O – SiO_2 for granites of north-eastern Africa.

Gattar granitic complex: 1—1st phase; 2—2nd phase; 3—3rd phase; 4 to 12—granites: 4—biotitic, Egypt (average of 28 samples), 5—riebeckitic, Egypt (average of 12 samples), 6—north-west Sudan, 7—Uganda, 8—batholithic, Egypt (average of 43 samples), 9—batholithic, north-eastern Sudan (average of 11 samples), 10—'young,' Nigeria (average of 7 samples), 11—'young,' Tarrauji (average of 3 samples), 12—rare metal, MPR; 13—Gujir rare metal granitic complex of Trans-Baikal (average of 11 samples); 14—alkaline granite of Dally; 15—alaskite of Dally. Field of magmatic series: I—alkaline, II—high aluminous, III—tholeiitic.

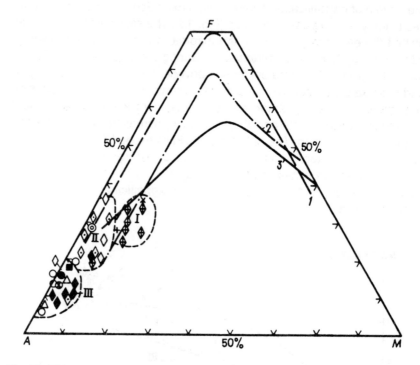

(66) Fig. 15. AFM diagram of granites of north-eastern Africa.

Line of differentiation: 1—Skeargard intrusive; 2—Hawaiian tholeiite; 3—calc-alkaline series. Field of gattar granitic complex: I—1st phase, II—2nd phase, III—3rd phase. Legend same as in Fig. 12.

(65)   The directional character of differentiation of the gattar granites from the early to the late phase is also distinctly reflected by the Kuno solidification index (M), which represents the ratio of magnesium oxide to the sum of oxides of magnesium, iron and alkalis. The value of this ratio for the majority of primary magmas is 40 and it approaches zero for fully differentiated residual melts. For example, the Kuno index for batholithic granites of the upper Riphean is 10 and for the gattar granites 5.45 to 5.28. The biotitic and riebeckitic granites of Egypt ($M = 0.52$ to 0.29) are the most differentiated varieties.

      Thus, the petrochemical features of the Vendian-lower Paleozoic post-tectonic granites help to include them under the differentiates of a deep-seated source of palingenetic granitoid magma, formed due to melting of the granitic and metamorphic layers. The relatively higher value of the ratio of isotopes $^{87}Sr/^{86}Sr$, which is 0.7201, indicates the 'crustal' level of their generation.

Table 4. Most important petrochemical characteristics of granites of north-eastern Africa and the Mongolian People's Republic (MPR).

Granitoid Complexes	K_2O/Na_2O	K_α	M	G
Gattar granites of Egypt				
Early Gattar	0.99	0.81	5.45	2.25
Late Gattar	1.11	0.92	5.28	2.41
Biotitic	1.10	0.83	0.52	2.09
Riebeckitic	1.15	0.91	0.29	1.99
'Young' Granite Massifs of NE Sudan				
Taggoti	1.08	0.57	8.51	1.53
Khalaib	0.48	0.90	1.38	2.18
El'ba	1.00	0.84	6.93	2.48
Gash-Amer	0.84	0.96	0.26	2.75
Umm-Shibrik	0.88	0.94	4.50	3.23
Ugandan Granites				
Lunio granites	0.45	0.98	0.18	3.41
Namaingo	1.26	0.68	0.24	2.10
Buteba	1.45	0.86	2.10	1.10
Mazaba	0.84	0.83	3.72	1.81
In-Tunina Akhaggara granites				
First phase	1.30	0.72	5.40	2.68
Second phase	1.29	0.78	1.78	2.63
Third phase	0.79	0.75	1.70	2.61
'Young' Granites of Nigeria				
Biotitic	1.17	0.93	0.19	2.35
Tarrauji Aira	1.09	0.89	0.61	2.84
Rare Metal Granites of MPR (Kovalenko and others)				
Standard Modotin type	1.12	0.80	2.00	0.28
Lithium-fluorine type	1.10	0.93	0.32	2.20
Albite-lepidolite type	0.58	0.90	0.90	3.72
Amazonite-albite type	1.00	0.32	0.60	2.40
Gujir complex	1.12	0.84	3.27	3.00

Note: $K_\alpha = K_2O + Na_2O/Al_2O_3$—Agpaiticity coefficient; $M = MgO/MgO + Fe_2O_3 + FeO + K_2O + Na_2O$—Kuno solidification index (%); $G = (Na_2O + K_2O)^2/SiO_2 - 43$—Ritman's index (%).

The problem regarding the age of the gattar or 'young' granites of north-east Africa and Arabia has yet to be thoroughly studied. This is obvious from the fact that granites of wide age range—from the late Riphean to the Cenozoic, are included in the 'young' group. The age of the gattar granites is more definite in Egypt, north-eastern Sudan and Saudi Arabia. There are, at present, more than 150 age data (K-Ar and Rb-Sr meth-

(66)

ods) for this territory, the majority falling within the interval of 650 to 450 million years, i.e., Vendian to early Palaeozoic. The gattar granites of Egypt are subdivided into three age groups—630, 610 and 570 million years. However, according to available data, the upper age limit of the granites should not be kept in the vicinity of 570 million years, insofar as the available radiological age determination data of granites from Egypt, Sudan and Saudi Arabia include many figures lesser than 570 million years, which correspond to the early Palaeozoic (Cambrian, Ordovician). K. Neary suggested an early Palaeozoic age of granites in north-eastern Sudan and separated the 'young' granitic phase having an age of 500 million years. These researchers distinguished granites with $\sim$ 100 million years age. This situation confirms the wide age range of the gattar granites.

(67) It should be noted in conclusion that the closeness in petrochemical composition and age data of the gattar granitic complexes encompassing the Vendian to lower Palaeozoic interval and even Mesozoic do not help to answer the question—where does the orogenic stage end and the activisation stage begin. An indirect answer to this question may be obtained on the basis of autonomy in tectonic localisation of the late gattar, particularly muscovitic and leucocratic granites, including them under the activised types and isolating them from the orogenic types associated with the concluding stages of development of the Red Sea late Proterozoic geosynclinal-fold belt. The age proximity of the Red Sea belt to the orogenic stage of development is also characteristic for magmatic complexes of the central type which also include gattar type granites. Thus, the activised development of the Red Sea rift zone in the prerift stage followed with a minimum hiatus after the orogenic period and resembled a 'trough' type. One characteristic feature of the Red Sea rift zone lies in its development at the prerift stage.

The concentric intrusives of the western boundary of the Red Sea are subdivided into two groups, namely the southern and northern, on the basis of their shape and composition. The intrusives of the southern group, Tekhilla, Takhamiyam, Haya, Kur, Sasa and Eight include derivatives of the mantle-originated tholeiitic as well as crustal granitoid and mixed magmas. These complexes, situated in the central part of the western boundary of Red Sea, are characterised by a predominance of alkaline-earth rocks over the alkaline rocks. Alkaline varieties predominate in the assemblage of rocks forming the northern group of concentric complexes which combine the intrusives, namely Umm-Shibrik, Uveinat, and also a few 'open' type concentric intrusives in the Egyptian part of the Red Sea hills (Abu-Khuruk, El Kahfa, Nugrus-El Fokani and others). The Ankur and Salala complexes occupy an intermediate place between these groups and they have both derivatives of tholeiite magma (troctolites or olivine gabbro) and also alkaline rocks (aegirine and nepheline syenites, essexites) in their composition.

Considering the age and extensive development of the platform cover over the territory of the northern group of complexes, such a distribution of the natural assemblages in concentric complexes leads to the suggestion that the alkali-gabbroid and gabbro-granitoid complexes of the southern Red Sea hills are the locales of earlier centres of mantle activisation in the prerift development stage of the Red Sea rift zone [28]. The concentric complexes of the northern alkaline group are, in general, more deep-seated according to the level of magma generation, much later in the period of their formation and less eroded, inasmuch as the majority of them include remnants of volcanic formations. The alkali-granitoid concentric complexes are developed more extensively than the first two groups. The empirically established pattern of emplacement of concentric complexes with regard to their relationship with rift genesis is examined below.

(68) The association of the concentric complexes shows that evolution from gabbro-granitoid to alkaline varieties took place gradually, and the shape of all the concentric complexes is almost stereotypical. The combination of morphological types indicates a generality in the development process of the complexes under discussion. This fact provides the basis for considering that all the concentric magmatic complexes depict, in their totality, different levels of magma generation in the evolution of the singular endogenetic system. Differences in the shape and composition of individual complexes depend on the level of erosion, but, on the whole, such do not exceed the limits of variations of the quantitative rock composition of the formation, which includes one or the other complex.

The direction and character of differentiation of the concentric intrusive rocks are represented in an AFM diagram (see Fig. 12). It can be seen from the diagram that the concentric intrusive rocks form an uninterrupted petrochemical series from the tholeiitic to alkaline varieties. With respect to $Na_2O + K_2O - SiO_2$ (see Fig. 13), the rocks, excepting the Sasa and Kur intrusives, belong to alkaline varieties. Their major part belongs to the sodic series and only the syenites of the Ankur and Kelly intrusives belong to the potassic variety.

Formation Period of the Central Type Magmatic Complexes

The above statements show that the central type magmatic complexes represent complex, heterogeneous and usually polyformational occurrences undergoing prolonged development. The formation of ring-like intrusives continued over an extensive period, starting from the late Proterozoic (Vendian) and extending up to the Cenozoic. Based on radiological (mainly K-Ar method) and geological data (in million years) the following epochs in their formation have been distinguished at present: 650, 550, 450, 400, 350, 230, 180, 140, 120 and 90, i.e., they were formed during the Vendian, Cambrian, Ordovician, Silurian, Devonian, Jurassic, Cretaceous and Palaeogene periods (V.I. Budanov, J.R. Vail, A.V. Razvalyaev, G.P. Shakhov and K.M. Serensits and others). They were not traced only in the Carboniferous and Triassic.

It is interesting to note that a prolonged period of formation of the central type complexes is not only observed in the Red Sea zone, but is also true for individual intrusives as well. Thus, according to radiometric age data (K-Ar and Rb-Sr methods), the durations of formation of the ring complexes for the intrusives are (in million years): Tekhila 636—493 Sasa 734—490, Salala 570—240, Ankur 250—121, Umm-Shibrik 180—140 and Abu-Khuruk 90—45. It follows from the aforesaid that the formation period of individual magmatic complexes varies from 35—50 million years (Abu-Khuruk) up to 330 million years (Salala). The formation of individual central type massifs is comparable with one or even a few tectonomagmatic cycles. Thus, for example, formation of the Salala ring complex started even in the Vendian (570 million years) through injection of alkali syenites of the external ring intrusive and continued further through the formation of the gabbro-anorthosite internal ring (540 million years) and (69) syenites of the central stock (430 million years). The formation of this ring complex was concluded by intrusion of nepheline-syenite stocks (240 million years). Considering that the Salala ring-complex is polycentric and the small ring complex is younger with respect to the main complex, it may be suggested that initiation of formation of the Salala complex took place in the Vendian or even in late Riphean.

The available radiometric age data of the rocks of ring magmatic complexes help to conclude that mainly the Vendian and early Palaeozoic intrusives are characterised by a prolonged period of development, encompassing 100 to 200 million years or more in the ancient period. The period of development of younger ring intrusives does not exceed a few tens of million years (50 to 60), which conforms with the duration of development of the Mesozoic and Cenozoic ring intrusives of Africa, Brazil, Canada and others. Moreover, the effect of tectogenetic impulses ranging from the early stages of tectonomagmatic cycles to the late stages is observed in the formation of the multiphase ring intrusives.

A definite pattern is observed in the distribution of ring intrusives when their age of formation is analysed. It has been established that the oldest, namely, the late Proterozoic (Vendian) and early Palaeozoic complexes are situated in the central part of the Red Sea framework. The ring intrusives of the Sudanese Red Sea hills mainly represent the given age group. Intrusives of this group are also developed on the opposite side, i.e., eastern part of the Red Sea as well as in Saudi Arabia. And although these are inadequately studied compared to those in Sudan, nonetheless, the available data do not contradict the accepted conclusion, but, on the contrary, confirm it, insofar as older age data apply to the concentric massifs situated in the central part of the eastern framework of the Red Sea. Moreover, numerous 'layered' gabbro intrusives are situated here, a part of which, compared with the Lakatakh intrusives, belong to central-type intrusives. The age of these massifs corresponds to the Vendian-early Palaeozoic range. All these data show that formation of the central type magmatic complexes of the late Riphean, Vendian and early Palaeozoic was localised in the central part of the Red Sea framework.

Younger (Chalk-Palaeogene) central-type intrusives are situated in the central part of the Red Sea in Egypt. These are ring complexes of El Mansura, El Gezira, El Naga, El Kahba, Nugrus-El Fokani and others. The range of age of their formation varies between 80 and 45 million years. In recent years, the occurrence of Cenozoic alkaline complexes, intruded into the Cretaceous nubian sandstones in the form of plugs and isometric stocks, has been established due west of the Nile River valley at the latitude of Aswan town. Apparently, the alkaline complexes of the ring-like and stock-shaped bodies in the Uveinat complex represent the continuation of the zone of these intrusives towards the west.

(70) Sequence of Formations of the Central Type Magmatic Complexes

It is known that magmatic formations, particularly the alkaline ones, are sensitive indicators of endogenetic regimes. In the last decade, formational analysis of magmatic rocks became one of the most important theoretical trends in geology and served as a fundamental basis for metallogenic correlations and theoretical formulations in the fields of geotectonics as applied to the creation of models of the deep-seated constitution and development of the earth's crust and the upper mantle.

The fundamentals of formational analysis of geological entities are incorporated in the publication of N.S. Shatskii, N.P. Kheraskov, Yu.A. Kuznetsov, Yu.M. Sheinmann and others. During the last decade, formational analysis has been used in many aspects of geology, such as petrology, metallogeny and tectonics.

The problems of sequential formation were reviewed for the first time with respect to the ring intrusives of the Arabian-Nubian alkaline provinces. Herein, analysis was based on the data available for concrete massifs, peculiarities of tectonic development of the region and theoretical concepts regarding the existence of initial magmas at different depths; the occurrence of these magmas under different tectonic set-ups gave rise to varied rock types. The main aspects of the concept of varied depth of magma generation are included in the publications of Yu.M. Sheinmann, Yu.A. Kuznetsov, L.S. Borodin, L.N. Kogarko, R.M. Yashina, V.A. Kononova, V.L. Masaitis, V.N. Moskaleva and V.G. Lazarenkov.

Insofar as central type intrusives consist predominantly of alkaline rocks, the problem of their sequential formation is related mainly to these formations. There are, at present, a few classifications for alkaline rock formations. At the same time, many problems of classification of alkaline rock formation are still debatable. There is in particular no unanimous view among petrologists regarding the nature of nepheline-syenite and alkali-syenite magmatism, and the occurrence of an independent nepheline-syenite magma is still controversial. Also, the nepheline-syenite rocks are regarded either as differentiation products of alkali-olivine-basalt magma, or are considered to have been formed due to the vertical shift of the magmatic source in the mantle with consequent generation of alkali-ultrabasic, alkali-basaltic and phonolitic magmas.

According to Yu.M. Sheinmann, all the alkaline rocks are subdivided into three formations: alkali-ultrabasic, gabbroid and alkali granitoid magmas corresponding to alkali-ultrabasic gabbroid and granitoid magmas respectively. Nepheline syenites, lying outside this scheme and unique in petrographic and petrochemical features, may be included in any of the three formations and thus be considered as differentiation products of the corresponding magma. L.S. Barodin differentiated the same three groups

(71) of alkaline rocks; however, he retained the same terminology for the first two, namely, alkali-ultrabasic and alkali-gabbro and suggested that the third group be named the alkali-granitoid (including nepheline syenites) formation (Yu. M. Sheinmann's concept). In his opinion, the latter has no genetic but only paragenetic interrelationship. According to L.S. Borodin's scheme, the nepheline syenites are the products of crustal melting of magma with participation of juvenile emanations. He holds that their spatial and chronological co-existence with granites is the result of parallel crustal melting.

The classification of alkaline rocks by V.G. Lazarenkov [16] is closely similar to that of L.S. Borodin's scheme. The former differentiated three formational groups: 1) alkaline ultramafites and gabbroids; 2) alkaline granites; and 3) nepheline and alkali syenites. The alkaline rocks are subdivided into 11 formations in the classification of V.G. Lazarenkov. The difference between his and L.S. Borodin's classification lies, first of all, in separation of an independent formational group of nepheline and alkali syenites, which, in his opinion, was formed (generated) at the mantle level. L.G. Kogarko also regards nepheline syenites as mantle-based in origin. In her opinion, the ubiquitous relationship of nepheline-syenite (agpaiticity) rocks with the parental deep-seated mantle-generated melts is indicated by such factors as spatial combination of alkali-ultrabasic and nepheline-syenite rocks, geochemical and petrochemical similarity, and also the presence of mantle-generated xenoliths in them. L.S. Kogarko noted that the relationship of nepheline syenites with the mantle attests to the 'dry' composition of the agpaitic magmas, their low oxidation potential, 'reduced' composition of gaseous phase and a number of other petrologic-geochemical features. As regards the Arabian-Nubian province, it is important to underline the relation of nepheline syenites with alkali-basaltoid formations, as observed by L.N. Kogarko. R.M. Yashina [33] suggested the independence of the nepheline-syenite magma and considers that nepheline syenites may become the differentiation products of alkali-basaltoid and alkali-ultrabasic magmas. However, in her opinion, large independent nepheline massifs of the type Lovozerskii, Khibin, Ilimaus-sak etc., are the products of independent nepheline-syenite (phonolitic) magma of mantle origin. D. Sutherland, L. Williams, N.A. Logachev and others concluded the existence of an independent phonolitic magma with the Kenyan rift of Eastern Africa as an example, where enormous, unique-dimensional plato-phonolites are not related to basaltic magma either in time or space.

The deep-seated petrologic aspect of formation of phonolitic magma poses a number of questions but their treatment is beyond the scope of the present investigation. It is only to be noted that, on the whole, after acknowledging the deep mantle level generation of phonolitic magmas, researchers differ in their assessment of depth, character of initial mate-(72) rial and magnitude of the process. Thus, in the opinion of L.G. Kogarko, agpaitic melts formed under prolonged differentiation of mantle-based melanephelinic magma in interaction with acidic volatile mineralisers. R.M. Yashina considers that the generation level of phonolitic magma was at $\sim$ 80 km depth, i.e., at a greater depth, than the generation level of olivine-basaltoid magma ($\sim$ 60 km), as theoretically estimated and ex-perimentally confirmed by D.H. Green and A.E. Ringwood. This has given rise to serious objections from a number of petrologists. The later con-cept of R.M. Yashina [33] is apparently more realistic, which assumes that phonolitic magma formed in the upper mantle in the frontal part of emerg-ing columns of alkali-basaltoid magma with the participation of fluids. As the process progressed the temperature fell, the alkali-basalt columns disappeared and an isolated source of phonolitic magma formed. Conse-quently, phonolitic magma is less deep-seated than the alkali-basaltoid magma. An analogous opinion is held by N.A. Logachev in explaining the origin of the plato-phonolites of the Kenyan rift.

Thus, the majority of researchers are inclined to support the concept of mantle-derived initial phonolitic magma for explaining the large phonolite areas or extensive independent nepheline-syenite massifs. As regards nepheline-syenite magmatism in continental rift zones where nepheline syenites occur in central type multi-phase polyformational complexes and are associated with alkali-ultrabasic and alkali-gabbroid formations, the conditions of such magma generation are less fully known due to inadequate study of this problem. According to the concepts of R.M. Yashina and other researchers, part of the nepheline-syenite massifs apparently generated as a crystallisation differentiation product of alkali basalt (gabbroid) magma. However, nepheline syenites and alkali syenites of such ring intrusives as Salala in the Red Sea rift zone, set apart from the formation of an alkali-gabbroid basic intrusive for not less than 300 million years, can hardly be the differentiates of the latter. It is more natural to consider them as derivatives of independent initial phonolitic magma.

Considering the aforesaid, the author regards the nepheline-syenite rocks as products of an independent mantle-generated nepheline-syenite (phonolitic) magma, formed at a higher level than the alkali-basaltic magma. This assumption conforms, on the whole, with the concepts of L.S. Borodin, R.M. Yashina and V.G. Lazarenkov.

It is important to take into account the position of the nepheline-syenite formation in the order of level of depth for magma generation while carrying out analysis on formations for the purpose of establishing the endogenetic regime of the Red Sea rift zone. The level of magma generation for alkali-basaltoid, alkali-ultrabasic with carbonatites and alkali-granitoid formations is presently theoretically established, experimentally confirmed and extensive petrologic literature is devoted to this subject in (73) the USSR and abroad. Some difficulties are encountered in determining the level of generation of phonolitic magmas. These may, however, be solved if we take into account the theoretical and experimental data of PT-conditions of formation of magmatic melts published by D.H. Green, A.E. Ringwood, V.S. and N.V. Sobolev, N.L. Dobretsova, L.S. Borodin, L.N. Kogarko, R.M. Yashina and others. We note on the basis of the foregoing publications that phonolitic melts are formed in the upper mantle at $P = 15 \cdot 10^8$ level, whereas alkali basalts are formed at great depths at $P = 25 \cdot 10^8$ level, alkali-ultrabasic melts with kimberlites at $P = 80 \cdot 10^8$ level and at a depth of 150–250 km.

In this book, the author supports the classifications of Yu.A. Kuznetsov, L.S. Borodin, R.M. Yashina, V.N. Moskaleva, V.G. Lazarenkov and others. They are more applicable and convenient for tectonic analysis, insofar as they correlate one or the other rock association with the initial magma generated at a definite level. Three main formational groups of alkaline rocks are distinguished among the central type intrusives in the Red Sea-Aden rift zone, such as alkali-gabbroid, nepheline-alkali-syenite and alkali granitoids. Rocks of the alkali-ultrabasic formational groups are developed to a very limited extent and are provisionally distinguished.

Chronological and Lateral Order of Formation of Central Type Magmatic Complexes

The concept of the order of chronological and lateral formation, displaying the evolution of the endogenetic regime, is the methodological basis for the analysis of formation of central type magmatic complexes, and the same has been published by N.A. Logachev, V.N. Moskaleva, V.G. Lazarenkov, Yu.G. Gatinskii and others. Recent investigations have shown that the paragenesis or chronological and lateral orders are more effective in different palaeoreconstructions of tectonomagmatic processes and not the individual magmatic formations.

The author accepts the chronological order of magmatic formations as the paragenesis of successively formed magmatic formations, and the lateral order as a synchronously formed large endogenetic system (in this case, the Red Sea rift zone) within adjacent structural components. The chronological order of magmatic formations of ring intrusives characterise the evolution of the endogenetic regimes with time, and the same are exhibited by the generation of magmas at different depths. Lateral orders display change in composition and structure of formations (or ring intrusives) which are more or less similar in age ($\sim$ 50 million years), and which are traceable at different parts of a large endogenetic structure, characterising the possible impulse of its magmatic activity. A comparison of lateral order shows the degree of development (advancement) of endogenetic regimes. The lateral-chronological sequence of one or the other large tectonic component, associated with the existence of a singular region of thermal stimulation in the mantle, is characterised by the assemblage of magmatic formations formed within the limits of the given endogenetic system.

74)   The composition of the central type magmatic complexes of the Arabian-Nubian province shows that the foremost among the complexes are the alkali-gabbroid and nepheline-alkali-syenitic types. These two formations constitute the base of the lateral-chronological order reflecting the direction of development of the endogenetic regime. In this two-member series, the alkali-gabbroid (gabbroid) formation is the older, the beginning of its formation was in the Vendian and, possibly, to the end of the late Riphean. As regards the alkali-granitoid formation, it seems to have a 'trough' development, insofar as ascent of deep-seated basaltoid and adjacent nepheline-syenite magmas somehow acted upon the sialic crust, leading to the genesis of palingenetic granitoid magma in it.

Analysis of the lateral and chronological order of the ring intrusives in the Red Sea rift zone enabled conclusions to be drawn on the character of development of the prerift endogenetic regime. First of all, the chronologic formational orders reveal the cyclicity of its development, which can be established from the change in the level of magma generation (Fig. 16). Three full cycles of change in the level of depth of magma generation have been distinguished in the evolution of the prerift endogenetic regime of the Red Sea rift zone. The duration of the first was 420 million years, and that of the second and third 80 and 70 million years respectively. The first cycle corresponds to the Vendian and Palaeozoic. The graph showing the level of magma generation during the first cycle is approximately symmetrical. The duration of time of the cycle reflects consistent activisation of the endogenetic regime at the level of the alkali-granitoid formation during the Cambrian, Ordovician, Silurian and Devonian. The (75) graphs for changes in the level in magma generation for the second and third cycles are approximately commensurate. In the time scale, these cycles correspond to the Mesozoic with maximum decrease of depth level of magma generation at around 180 to 120 million years.

Analysis of the chronologic-lateral formational orders of the central type magmatic complexes revealed a number of peculiarities in the de-

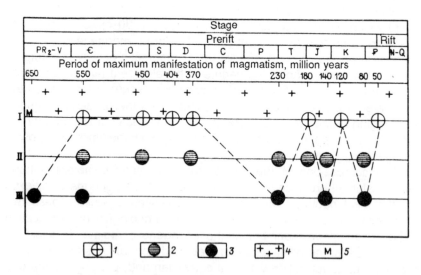

(74) Fig. 16. Generalised lateral-chronological order of formation of ring complexes of the Red Sea rift zone.

Magma generation sources for: 1—alkaline granites; 2—nepheline and alkali-syenites; 3—alkali-gabbro; 4—earth's crust; 5—Mohorovičić boundary. Levels of magma generation: I and II—$15 \cdot 10^8$ level; III—$25 \cdot 10^8$ level.

velopment of the Red Sea rift zone in the prerift stage, which crosses the boundaries of the regional values. These are: 1) confinement of maximum endogenetic rejuvenation to the Precambrian-Phanerozoic boundary ($\sim$ 550 million years), 2) confinement to the Palaeozoic-Mesozoic boundary with sharp change in the level of magma generation, reflecting an increased contrast of endogenetic regime in the Mesozoic and Cenozoic.

Thus, magmatism in the Red Sea rift zone, along with its manifestations of increased activity in particular epochs, was marked by major cyclicity, interrelating chronologically with the global datum-line of geologic history, such as Precambrian-Palaeozoic, Palaeozoic-Mesozoic boundaries or the epoch boundaries, which are distinctly separated by significant reconstruction of geologic history, accompanied by global transgression of seas and development phases of newly formed oceans (late Jurassic and late Cretaceous).

The investigations carried out so far enabled quantitative evaluation of the endogenetic regime in the prerift stage. This is revealed by the number of age boundaries in the manifestation of magma generation. In the order of increasing depth level of magma generation for the Red Sea rift zone the values are as follows: for Levels I-7, II-7, III-5, IV-0, or these manifestations may be represented in the form of ratios 7:7:5:0; the same values for the Baikal region are 3:3:1:0 (see Fig. 16). The suggested form of quantitative assessment of endogenetic regimes gives an idea of the quantitative interrelationship of the age limits and levels of magma generation. And although this approach does not contain information on the dimension of magmatism, it may still be considered as the first step on the way to formulating the methods for comparable assessment of prerift endogenetic regimes.

Structural-Tectonic Conditions of Localisation of Central Type Magmatic Complexes

An analysis of the spatial disposition of ring intrusives in the Arabian-Nubian alkaline provinces showed that alkali magmatism displays a distinct tectonic control. The central type magmatic complexes are confined to rift zones which may be regarded as 'penetrative fault structures' according to Yu.M. Sheinmann or lineaments according to L.S. Borodin. The relation of alkaline ring intrusives with major faults, including rift faults and also their converging disposition within rift zones is well known. In this regard, data on the disposition of concentric intrusives of the Arabian-Nubian alkaline province are not exclusive, but rather provide a new (76) concept regarding the relation of ring intrusives with rift genesis both from a structural as well as historical-geologic aspect.

The ring intrusives in the Arabian-Nubian province are controlled by faults of north-west (Red Sea), north-east and sublatitudinal (Mediterranean) orientation that are characteristic for this region of Africa. El'Ramli, V.I. Budanov, F. Delani, J.R. Vail, A.V. Razvalyaev and K.M. Serensits indicated such a relationship of concentric intrusives with these faults. However, data on the Red Sea hills show that major submeridional zones of deep-seated faults with a prolonged period of development played a significant role in the disposition of ring intrusives.

The interrelationship of ring intrusives with submeridional fault zones has considerably broadened the concepts regarding the constitution of the Red Sea rift zone, history of its formation and interrelationship with the East African rift system. It had been established that the faults of this strike play a leading role in the constitution of the Red Sea rift. Submeridional faults are more often concealed and discontinuous in nature and therefore may be identified using a combination of direct and indirect geologic, geomorphologic and geophysical indications. Two such zones, namely, Diib and Barak have been studied in the Red Sea hills (see Fig. 8).

Major submeridional faults are the magma feeders for alkali magmatism. These faults were formed prior to the genesis. As a result, the manifestations of alkali magmatism became emplaced in the pre-existing structural set-up during the development of the Red Sea rift. The alkaline ring intrusives in the Red Sea zone are grouped as linear extended zones (belts) with a north-westerly strike. With thickness varying from 50 to 150 km, these zones occur bordering the rift valleys and are traced parallel to them. Zones of alkali metasomatism reveal a genetic commonality with the Red Sea rift and are regarded as unique features of the pre-rift stage of epiplatform activisation, such as activisation furrows, which break up the platform into banded blocks activised to varying degrees with a general north-west strike. The coincidence of the configuration of these belts with the structural plan of the Red Sea rift is of specific importance.

The relationship of ring intrusives with the structural set-up of the Red Sea rift has not only been displayed in the parallelism of the ring-like intrusive belts and structural set-up of the rift zones, such as the main and axial 'troughs' and bordering 'shoulders,' but also on the intimate relationship with the structural framework of rift genesis. In this regard, the relationship of ring intrusives with transverse faults of the rift is interesting. It has been established that the ring intrusives are grouped into separate linear zones with a north-east strike in the Red Sea structural framework. Faults have been detected in this strike all along the western boundary (77) of the Red Sea from Egypt in the north up to Eritrea in the south on the basis of a series of direct and indirect indicators. Geologic-geophysical investigations conducted in later years with the participation of the author in the Red Sea hills of Sudan established that the north-east-striking faults are of ancient, late Riphean-Vendian origin and their development took place in the orogenic stage of the late Proterozoic geosynclines.

Analysis of the lateral orders of magmatic formation in the ring complexes on a much wider plan beyond the framework of fault dislocations control showed that a definite pattern is discernible in their distribution according to composition. Alkali-gabbroid intrusive formations, differentiated intrusive formations of basic composition and tholeiitic dykes are developed in the central part of the Red Sea framework. Also they are the oldest in age. Nepheline-alkali-syenitic and alkali-granitoid intrusives occupy predominantly the peripheral position in the Red Sea rift zone. Accordingly, as one proceeds from the centre of the Red Sea rift zone to its periphery, a change in level of magma generation is observed, as though expansion of the endogenetic regime occurred in time and space (Figs. 17, 18).

Based on directional change in the formational composition of central type magmatic complexes in the western Arabian-Nubian province and, consequently, on the change of level of generation from the centre of the Red Sea rift to its periphery, a conclusion may be drawn about the existence of an activised zone in the prerift development stage of the Red Sea rift. This activisation zone is oval in shape in plan and arose out of the earliest sources of prerift stimulation of the mantle. It is important to note that the Red Sea rift coincides with the long axis of the oval-shaped zone of activisation. (see Fig. 8).

As regards the eastern flange of the zone of activisation, it may be observed that although it has been inadequately studied in the Saudi Arabian territory, still the available data on the tendency towards changing depth of the hearth of the ring intrusives due east of the Red Sea may be contemplated. Firstly, the alkali-gabbroid ring intrusives developed here and it is to be specially noted that such intrusives occurred in combination with layered gabbro-norite-anorthosite massifs of late Precambrian-early Palaeozoic age. Finally, all these intrusives localised within the Red Sea framework. The occurrence of a few ring intrusives of gabbroid composition in the extreme eastern part of the Arabian-Nubian province cannot offset the observed pattern and they may apparently be regarded as an outcome of subsidiary thermal stimulation of the mantle that occurred simultaneously with the main activisation zone of the Red Sea rift. It was earlier observed by the author that it would be logical to link the development of alkali-basaltoid volcanism of the Neogene-Quaternary period in the Druz trough of south-west Syria with the north-western continuation of the aforesaid zone. The fact that these two prerift and solely rift activisation regions occur as disconnected ones should be of no consequence for considering them interrelated insofar as rift genesis is characterised

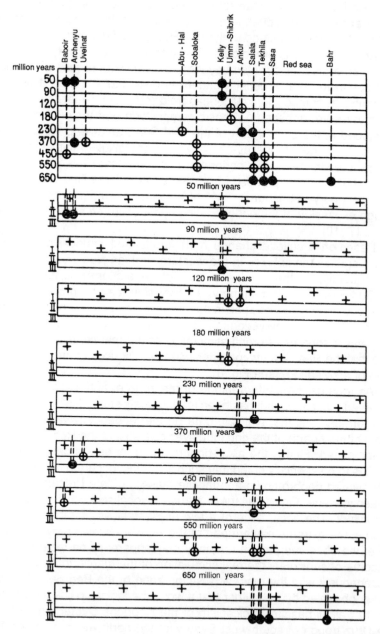

(78) Fig. 17. Lateral-chronological order of formation of ring complexes of the Red Sea rift zone (across strike).

Legend same as in Fig. 16.

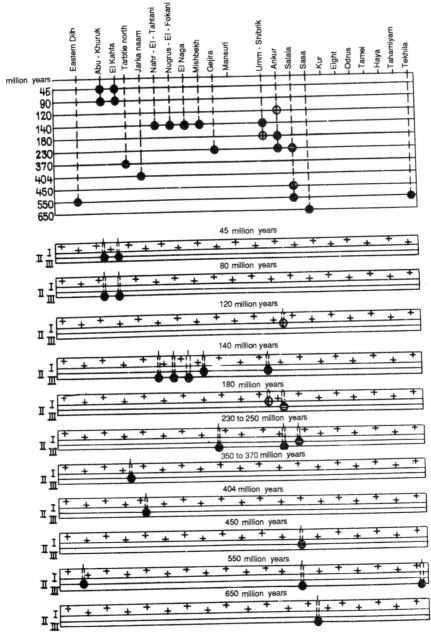

(79) Fig. 18. Lateral-chronological order of formation of ring complexes of the Red Sea rift zone (along strike).

Legend same as in Fig. 16.

(80) by discreteness in strike and, consequently, individual segments of the rift zones may occur in different stages of their development.

The structural localisation of the Druz volcanogenic region becomes clear in the light of the aforesaid statements. Attempts to link the tectonic set-up of this volcanogenic region with the Levantin fault zones remain unclear in the light of traditional approaches. Following the concept of laevo-shift of faults, the Druz volcanic region, having a general north-westerly strike of magma controlling faults, must have been affected by uncompressive stress even if it is a component of horizontal nature. Such a kinematic state does not favour the emergence of volcanism and, consequently, cannot explain the nature of this major volcanic region, which is diagonally oriented to the faults of the Dead Sea rift. Therefore, the relationship of magma controlling faults of the Druz region with the north-western structural set-up of the Red Sea rift genesis zone is more natural. Interestingly in recent years north-easterly striking faults (Red Sea type) have been revealed by cosmic photographs in north-western Syria. These data significantly widen the significance of the Red Sea structural trend in the constitution of the Arabian plate, as recently confirmed by G.D. Azhgirei and others.

The southern terminal of the prerift activisation zone of Red Sea rift zone is inadequately delineated. The southernmost ring intrusives of alkali-gabbroid composition are situated at 18° N lat. further south, in Eritrea. It is still inadequately studied and naturally information on the ring intrusives of this region is fragmentary. A major part of Ethiopia is covered by plateau basalts of the Eocene-Neogene trap series. Ring intrusives are delineated in the northern and western boundaries of plateau-basalts. To the north, in southern Eritrea, this is represented by the Gemalha gabbro-granitoid complex associated with rhyolites and agglomerates. In western Ethiopia, the ring intrusives are represented by three gabbroid massifs; the largest among them is the Duma intrusive and the other two the Duma and Gangan syenitic ring intrusives. It is important to note that the syenitic intrusives also occupy a peripheral position among these intrusives with respect to the rift, in this particular case the Ethiopian rift, i.e., the same tendency as seen in the Red Sea rift zone is maintained in their disposition.

In concluding the analysis on magmatism during the prerift development stage in the Red Sea zone, it is necessary to bring out the important new positions attained which are significant in understanding the rift genesis process.

1. An independent Arabian-Nubian alkaline province has been distinguished. The multiphase nature of its development has been established here for the first time and spatial-chronologic characteristics have been enumerated.

2. It has been established that the Arabian-Nubian Alkaline Provinces had a prolonged period of development. Magmatism continued from the Vendian to the Cenozoic. The beginning of activised magmatism essentially coincides with magmatism of the orogenic stage of development of the late Proterozoic Red Sea mobile belt. Similarly, almost 'penetrative' development of the prerift activised magmatism is a unique feature of the Red Sea rift zone.

(81)

3. Multistage development is characteristic not only for alkaline provinces on the whole, but also for individual magmatic complexes of the central type; the age range of such formations varies from 50 to 300 million years.

4. Chronologic-lateral orders of magmatic formation of the prerift stage have been detected and their cyclicity has been shown. Three full cycles of change in magma generation level have been traced in the evolution of the prerift endogenetic regime of the Red Sea rift zone. The duration of either of these cycles or for alkali magmatism epochs reduced with decreasing age, but the number of alkali magmatism epochs increased; this confirms intensification of the tectonomagmatic process in the earth's evolution.

5. Local and regional tectonic controls of alkali magmatism have been established. The first is displayed in confinement of central type magmatic complexes to predominantly submeridional or north-easterly striking faults, and, to a lesser extent, to latitudinal and north-easterly, and also to intersection regions of these faults; the second control is displayed in its localisation in an oval-shaped (in plan) zone of activisation which coincides with the Red Sea zone. Prolonged and consistent development of this zone led to its recognition as an 'activisation core' or 'riftogenic core'.

6. Lateral formational inhomogeneity of the activised zone is exhibited in the localisation of ring intrusives of gabbro-granitoid or alkali-gabbroid composition, 'layered' gabbro and a tholeiitic dyke in its centre, and nepheline-alkali-syenite and alkali-granitoid rocks at the periphery. Also the intrusives situated at the centre are more ancient (Vendian-Early Palaeozoic) and those along the peripheries are younger. It has been concluded on this basis that endogenetic activity expanded in the prerift stage of development of the Red Sea rift zone.

7. It has been established that the characteristic feature of magmatism in the prerift development stage of the Red Sea rift zone lies in multiple alternation of mantle (alkali-gabbroid, nepheline-alkali-syenitic and tholeiitic-basalt formations) and crustal (alkali-granitoid formations) magmas and, as a result, an assumption has been made regarding the thermal 'instability' or 'destabilisation' of the lithosphere.

3

Main Features of the Tectonics and Developmental History of the Red Sea-Aden Rift Zones

(82) The African-Arabian rift belt is one of the largest among the continental rift belts of the world. It extends for more than 6000 km from the Alpine folded belt in the north up to Mozambique in the south. With a general submeridional strike, its individual members and branches diverge to the north-west or north-east. Besides significant extension, the distinguishing characteristic feature of the African-Arabian rift belt with regard to other belts or rift zones appears to be its link with the mid-oceanic rifts of the Indian Ocean through the Aden rift. Individual rift zones within the rift belt show changes in their formation from typically continental (Nyaskan, Tanganyikan, Kenyan, Ethiopian) to intercontinental (Red Sea, Aden) types. In the former group, destruction of the continental crust took place in the initial stage, while in the latter, this process continued to the stages of its considerable transformation to the extent of replacement by newly formed oceanic type crust.

Voluminous literature, containing results of many researches on individual zones as well as on the African-Arabian belt, as a whole, have been published since the discovery of the Kenyan rift or Gregory rift by the English geologist John Gregory. All these publications are devoted to diverse problems of geology (stratigraphy, volcanism, geomorphology, geophysics and tectonics) of the aforesaid zones and the belt. Mention is made here of only the more important studies, which include research on the Kenyan rift by John Gregory, V. Baker, E. Saggerson and L. Williams; on the Ethiopian rift and Afara rifts by G. Taziev, P. Moore, J. Barry, R. Barberry and many others; on the Aden rift by A. Azaroli, E. Lauton and others; on the Red Sea rift by S. Tromp, D. Schwartz, D. Arden, A. Bateman and others; on the Western Arabian (Levantin) rift belt by L. Duberter, L. Picker, B. Willis, A. Kennell, Ya. Bentor, R. Freund and others; and on the whole African-Arabian rift belt or its individual large segments by F. Dicksie, R. McConnell and others.

Researches by the International Scientific Expeditions for solving major problems of the African-Arabian rift belt and also on the rift formation, as a whole, have contributed greatly to the study of the African-Arabian rift belt. The first were contributed by the Soviet East African Expedition (1967–1969) under the leadership of V.V. Belousov, in which the rift zones of Eastern Africa were studied; researches by the Italian and German (FRG) geologists in the Afar (1975–1976), and others, and also oceanologic investigations in the Red Sea and Aden Bay by the scientific expedition ships 'Wanda River' (1969) and 'Waldivia' (1971–1972) of the FRG. The investigations carried out in 1979–1980 (83) by the Red Sea Expedition of the Oceanology Institute, Academy of Sciences, USSR with the help of scientific expedition ships (NIS) 'Academic Kurchatov', 'Professor Stockman' and 'Aquanaut' and underwater apparatus, including the manned apparatus 'Paisis', should be specifically mentioned.

Considering that a résumé of the systematised regional data on the African-Arabian rift belt has already been published, and acknowledging the fact that the monographs of E.E. Milanovskii, A.F. Grachev and N.A. Logachev are more complete with respect to the northern megasegment, the author refrains from undertaking the task of detailed presentation of the extensive regional data on the African-Arabian rift belt.

The delineation of ancient as well as newer structural frameworks and the developmental history of the northern part of the African-Arabian rift belt in connection with the problem of predetermination of continental rift genesis is the particular objective of this chapter. In this regard, the Red Sea rift (Fig. 19) represents a good subject insofar as it combines in itself structures in which prerift development stages may be traced, on the one hand, within the Precambrian crystalline shield with excellent preservation of the magmatic formations of the activisation stages (Red Sea rift), and on the other hand, within the platform cover recorded in the sections of sedimentary rocks, their facies and thicknesses (Suez rift, Western Arabian or Levantin rift zones).

Red Sea Rift Zone

FORMATION OF THE FOUNDATION ('FRAME') OF THE RED SEA RIFT ZONE

The type of correlation of the riftogenic structural set-up with those of the prerift stage acts as one of the aspects for the predetermination problems in continental rift formations. The solution of this problem includes analysis of the contribution of the basement or 'frame' in which the Red Sea rift zone was formed. Insofar as the Red Sea rift zone was formed mainly on the Precambrian basement and its main features of constitution and development were reviewed in Chapter I, this chapter is devoted only to those characteristics of its formation that are necessary for revealing the nature of correlation among the prerift and rift stages.

It was observed in Chapter I that the Red Sea rift depression is intersecting (discordant) with respect to the Precambrian structural set-up. In this respect three sectors may be identified within it based on the character of correlation with Precambrian structures: central, northern and southern. Maximum discordant relationship of the depression with the Precambrian structural set-up is characteristic for the central sector; it is less sharp in the northern and southern (see Fig. 2). In connection with such sharp isolated correlations of the depression with the basement, its (85) formation and relationship with the riftogenic structural set-up, particularly in the central sector, i.e., in the region with extreme discordant correlation might be considered. Such an approach is dictated by expediency also, given the fact that the western boundary of the central part of the Red Sea (Red Sea hills of Sudan and Egypt) has been better studied by geologic-geophysical methods in recent years.

The main structural component of the foundation of the Red Sea hills is the Sudanese-Arabian late Proterozoic fold belt; three structural compositional complexes took part in its formation: Katarchean-lower Proterozoic, upper Proterozoic (Riphean) and late Riphean-Vendian (Fig. 20).

The volcanogenic-sedimentary green-schist complex of the lower-middle Riphean, represented by andesites, basalts, rhyolites, schists, greywackes, tuffs, conglomerates and marbles, are mainly developed within the Sudanese-Arabian fold belt. The complex shows folded forms of varied morphology with a consistent north-eastern strike. In magnetic characteristics, the rocks constituting the sedimentary-volcanogenic green-schist complex belong to the non-magnetic and weakly magnetic category, having magnetic susceptibility of up to $500 \cdot 10^{-6}$ international system of units. However, extensively developed intrusive formations within the complex with higher magnetic properties (up to $5000 \cdot 10^{-6}$

---

(85) Fig. 19. Scheme of structural components of the Red Sea rift zone.

1—Precambrian basement; 2—Phanerozoic platform cover; 3—riftogenic complexes (a—volcanogenic, b—sedimentary); 4—riftogenic faults (a—established; b—assumed); 5—regional faults (a—established; b—assumed); 6—faults established from geophysical data; 7—zones of magnetic anomaly (a—positive; b—negative); I—South Egypt, II—Nugruss, III—Halaib, IV—Dungunab, V—Mohamedkol, VI—Port Sudan, VII—Sinkat, VIII—Derudeb, IX—Karor; 8—gravitational stages (a—within the continent, b—assumed in water areas); A—Gebeit, B—Meder, C—Inderaikvan, D—Kerian, E—Kassalin; 9—deep-water basins with metalliferous muds and thermal solutions; 10—magmatic ring complexes; 11—volcanoes; 12—epicentres of earthquakes.

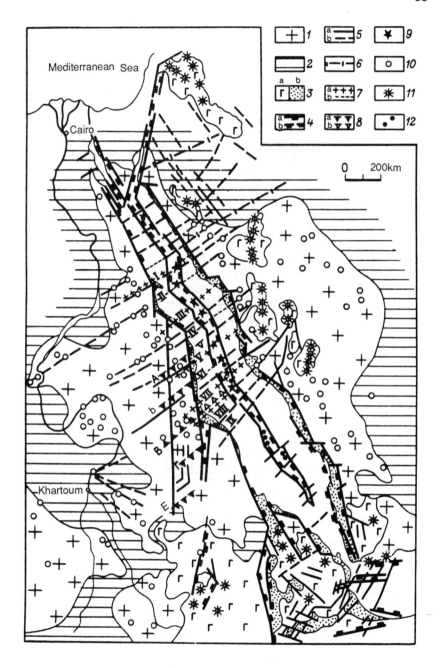

84

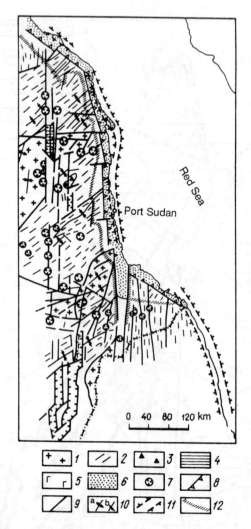

(86) Fig. 20. Tectonic scheme of Red Sea hills (according to E.N. Isaev and A.V. Razvalyaev).

Structural complexes: 1—Archean-lower Proterozoic schist-gneiss (Kashebib series); 2—upper Proterozoic volcanogenic sedimentary (Nafirdeib series); 3—late Riphean-Vendian orogenic complex (Avat series); 4—platform cover ('Nubian sandstones'); 5—Cenozoic basalts; 6—Cenozoic deposits; 7—ring intrusions; 8—marginal faults of Red Sea basin; 9—major faults; 10—axes (a—anticlinoria, b—synclinoria); 11—precoastal horst; 12—precoastal gravitational step.

international system of units) give a mozaic-like picture of the intensively differentiated magnetic field.

The youngest late Riphean-Vendian complex of Precambrian rocks of the Red Sea hills corresponds to the orogenic stage of development of the late Proterozoic Red Sea fold belt. It is represented by molasse strata of very limited occurrence. The complex consists of conglomerates, shales, dacites, andesites, rhyolites and their tuffs. The rocks are weakly metamorphosed (not above epidote-chlorite facies) and have almost a horizontal slope; they constitute low-dipping trough-like depressions and narrow prefault troughs, with increasing dislocation intensity in the vicinity of their boundary faults.

(86) The intrusive formations of the basement in the Red Sea rift zone are varied and represented by the late Proterozoic ultrabasites, gabbro, diorites and calc-alkaline batholithic granites. Rocks of basic and ultrabasic composition occurring in the form of small intrusions and dyke bodies belong to the early intrusive complexes. The batholithic granitic complex (granites, granodiorites, rarely diorites), covering more than 50% area in the Sudanese-Arabian belt (Red Sea hills) sharply predominate in quantity among the intrusive rocks. An extensive group of granites with characteristic multiphase and alkaline composition have been isolated among the orogenic stage intrusives of the Sudanese-Arabian belt. The central type intrusives are extensively represented; their period of formation corresponds to late Riphean to Mesozoic and even Cenozoic.

The structural set-up of the Sudanese-Arabian belt is characterised by a general north-eastern strike of folds and faults and magmatic sources. It is also manifested in large linear negative residual gravitational anomalies related to the schistose and gneissic complex in the intensively differentiated zonal magnetic field and in the development of intrusive belts. The main components of the Sudanese-Arabian structure are the zone of synclinoria and anticlinoria. The most important zones of synclinoria (from the north to the south) are: Daraheib, Dungunab, Port Sudan and less distinctly revealed (fragmentary) Tahamiyam and Langeb, separated by major batholithic granite massifs intruded into Halaib, Mohamedkol, Sinkat and Derudeb anticlinorial zones.

(87) Besides the general north-eastern direction, individual fragments of anomalous zones having a north-eastern strike are significantly manifested in the magnetic field of the Red Sea belts. Precambrian intrusive bodies, extending meridionally and thereby conforming to gravitational data, are also identified. Latitudinal strikes are observed in a few zones of intensive dyke magmatism, controlled by magnetic anomalies.

North-eastern-striking Precambrian structures are mapped up to the boundary of the Neogene-Quaternary deposits of the precoastal plains. E.N. Isaev established on the basis of the nature of magnetic fields that they may also be traced to the north-east from this boundary and are sharply interrupted by a complex system of faults comprising intersecting dislocations with predominantly a meridional and north-west strike forming the south-west side of the Red Sea rift depression. The differentiated north-east-striking magnetic field within the Red Sea has been established only in the extension of the Amarar zone in the region of 21°15′ N latitude.

The differentiated magnetic field forms a complex system of north-easterly striking subparallel zones in the Red Sea hills territory. There are five such zones, namely, Sofai, Amarar, Port Sudan, Sinkat and Derudeb (Fig. 21). The zones of differentiated magnetic field are subdivided by a major minima in the Bouguer anomaly conforming to the schist-gneissic complex. Gravitational stages with a north-east strike are distinguished within the minima or at their boundaries at Hebeit, Meder, Inderaikvan, Kerian and Kassalin. The minima of the Bouguer anomaly, regions of differentiated magnetic field and gravitational stages are in full conformity with the strike of the Precambrian Sudanese-Arabian fold belts. Also, the geophysical anomalies coincide with the known anticlinoria and synclinoria only in rare cases. According to the author, this fact is evidence of the post-fold (orogenic) stage of structural reconstruction during which the block structure of the region was formed. During this stage, individual large blocks, commensurate with regions distinguished from geophysical data, underwent a multidirectional vertical shift along their bordering weak zones (faults). Possibly, the north-east-striking gravitational fields and boundaries of differentiated magnetic field zones are mainly associated with such zones. Their coincidence with major interblock fault zones speaks in favour of such a postulate. The nature of the Hebeit fault, coinciding with the southern boundary of the Sofai zone of differentiated magnetic field, may be cited as an example.

The Hebeit fault is distinctly manifested in the sedimentary-volcanogenic rocks of the Nafirdeib series by the formation of schistosity on a large scale. Linear extensive bodies of quartz porphyry, regarded as effusive analogues of orogenic granites, are confined to them. The formation of schistosity in rocks is a process superposed on the folded structure of the belt. Consequently, schistosity together with the associated faults formed in the post-fold stage. Faulting took place not later than the orogenic stage (late Riphean-Vendian). This has been established from the fact that extensive bodies of quartz-porphyry are confined to faulting and also from its association with the late orogenic intrusive alkaline granites. These alkaline granites intersect the Hebeit fault without dislocating it. The age of the granites is 470 million years according to the K/Ar method. However, these values are somewhat lower since analogous granites in adjacent parts of Egypt date 600 million years by the Rb-Sr method.

Thus, the structural set-up of the Precambrian Red Sea hills of Sudan, forming the 'framework' of the rift zone, is worked out from the general north-east strike of the structural elements formed during the folding stage of early-middle Riphean intracratonic troughs. The orogenic stage (late

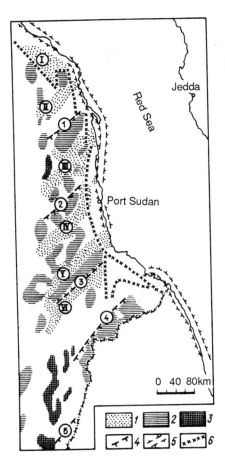

Fig. 21. Region-wise break up of Red Sea hill territory on the basis of geophysical data (after E.N. Isaev).

1—zone of differential magnetic field: I—Halaib, II—Sofai, III—Amarar, IV—Port Sudan, V—Sinkat, VI—Derudeb; 2—regions with negative residual gravitational anomalies; 3—Bouguer negative anomalies associated with grabens; 4—gravitational stages of small amplitude: 1—Gebeit, 2—Meder, 3—gnderaikvan, 4—Kerian, 5—Kassalin; 5—boundaries of near-coastal gravitational maximum; 6—precoastal gravitational stage of large amplitude.

Riphean-Vendian) is characterised by breaking up of the territory into north-east-striking blocks, separated by gravitational stages which coincide with major interblock deep-seated faults.

CENOZOIC (RIFTOGENIC) STRUCTURAL SET-UP OF THE RED SEA RIFT ZONES

The largest intercontinental rift zone of the Red Sea comprises wide depressions formed by grabens having a length of 2000 km approximately and a width ranging from 150 km in the north and up to 350 km in the

(89) south. The depression appears to be 'split up' in the north; the Suez Bay graben serves as its direct continuation and the narrow Aquaba Bay joins with it at an angle. In the south, the Red Sea depression narrows down at approximately 12°30′ N latitude and tapers out in the Bab-el-Mandeb strait, joining it with Aden Bay.

Newer uplifts comprising Precambrian rocks border the depression from the west and east. The contemporary rift zone structure is asymmetric in shape both in transverse and longitudinal sections in the 0.5 km interval structural contour map of the pre-Miocene surface. The uplift bordering the depressions on the west (Nubian) represents a narrow semi-arch 100 km wide in the north and up to 300 km in the south. It attains 1 to 1.5 km in height within the Arabian desert (Egypt) in the north and the same increases up to 2.5 km in the south. The eastern (Arabian uplift) boundary of the Red Sea depression is considerably wider than the Nubian depression. Its width is 300 to 450 km in the 1 km interval structural contour map and attains 450 to 700 km in the 0.5 km interval. The height of the uplift attains 3.0 km in the south. The vast volcanic plateau with axial volcanic ridges oriented NNW and submeridionally (Rahat, Kisb, and other plateaus) play a significant role in the formation of the eastern (Arabian) uplift. Both in the east as well as in the west, the southern ends of the bordering uplifts are covered by vast basaltic regions (Trap Series) of Eocene-Neogene age. The Arabian and the Nubian uplifts are complicated by arch-block uplifts and graben-form depressions of the second order.

The general structural plan of the Red Sea rift zone is characterised by a zonal banded structure. A main depression (main trough) represented by a graben separated from the 'shoulders' or 'wings' of the uplifted blocks of the rift by normal faults, is distinguished in the Red Sea rift structure. An axial trough is situated in the central part of the main depression. This corresponds to deeper water regions of the main sea and can be traced along the 1000 m isobath (depth contour).The axial trough is 50 to 60 km wide. According to the data provided by A.S. Monin, L.P. Zonenshain and others, the following large structural components are distinguished along the transverse profile of the Red Sea rift at 18°00′ N lat.:

1) Shelf zone, or upper stage with a width of up to 70 km and depth extending up to 600 m; 2) hillocky, terrace-type main scarp with 600 to 800 m height and 40–50 km width; 3) axial trough with 35 to 40 km width and characterised by highly dissected relief of the bottom having the maximum depth (up to 2000 m and more). Internal zones or lower tectonic stages, in turn, are distinguished (from one to three) at depths varying from 1100 to 1500 m in the axial trough and marked by a 4 to 5 km wide central uplift in the axial region.

The symmetrically situated structural stages bordering the axial trough geomorphologically represent shelf zones. Their width varies from 30 km in the northern part of the Red Sea to about 150 km in the southern part with relatively shallow depth (up to 500 m). Externally, the shelf steps are usually bordered by coastal steps adjoining the marginal faults of the rift. At 20° N lat. the trough is 100 to 150 m wide. An axial trough, attaining up to 2000 m width, separates within the main trough further south of 24° N lat. The axial trough is dissected at places by longitudinal furrows. Further south of 19° N lat. the depth and width of the main trough decrease with simultaneous widening of the shelf zone. The axial trough dies out at 17° N lat.

(90)

The majority of researchers consider that the main trough of the Red Sea rift represents a graben, filled with Cenozoic sediments, exposed only in the coastal plains (stages). Seismic investigations established that the foundation of the main trough is buried at a depth of 3—5 km (Fig. 22). The Neogene sediments exposed in the littoral plains, along with the entire coasts of Egypt, Sudan and Ethiopia, comprise terrigenous and terrigenous-carbonate rocks of lower and middle Miocene, evaporite strata of upper Miocene and coarse-fragmentary continental and littoral marine terrigenous-carbonate deposits of Pliocene-Quaternary age. The sediments attain a few kilometres in thickness, reaching 2.0—2.5 km in the Sudanese section of the Red Sea coast. The sediment thickness encountered in boreholes for oil drilling in the water areas off the Sudanese and Ethiopian coasts increases up to 3.8 km and more, and, according to seismic data, attains 5 km.

(91)

The Neogene sediments of the coastal steps of Egypt and Sudan are characterised by sharp changes in thickness and facies which are associated with horst-block formation of the transition zone from the rift-uplift bordering areas ('shoulder') to the shelf zone of the main Red Sea trough. The formation of this depression was accompanied by active development of parasedimentary faults and further development of uplifts and troughs which had been inclusively filled with rapidly accumulating detrital material. The active tectonic situation in the coastal zone of the main trough favoured basaltic eruption in coastal Sudan and Saudi Arabia.

The constitution of the shelf steps of the main Red Sea trough were studied by geophysical methods and exploratory wells. Seismologic investigations established that the reflecting horizon S, identified from the top of the Miocene evaporite strata, extends under the entire depression (excluding the axial trough). This horizon is identified as the surface of unconformity between the late Miocene evaporites and the weakly lithified Pliocene-Quaternary sediments deposited on them. The velocity of longitudinal waves in the evaporite rocks is 4.4 km/second and 1.6 to 1.7 km/second in the Pliocene-Quaternary sediments.

90

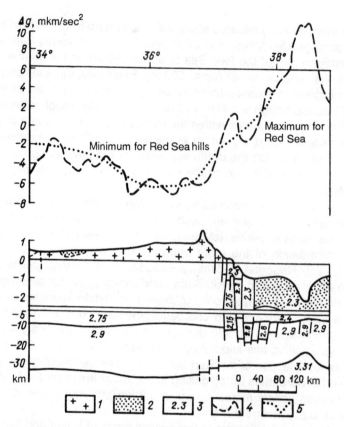

Fig. 22. Section of the crust in the eastern border of the Red Sea rift (after E.N. Isaev and A.V. Razvalyaev).

1—Precambrian substrata; 2—sedimentary cover; 3—density, g/cm³, according to gravimetric data; 4—Bouguer anomaly; 5—weighted average of Bouguer anomaly.

Seismic data show that Neogene deposits of the shelf steps in the main trough are dislocated by numerous faults and shear fractures have been observed in individual cases. Such have also been established from intensive heat flow, i.e., volcanism. The characteristic structural feature of the Neogene deposits of the Red Sea basin is the extensive development of diapiric salt structures and associated fold dislocations.

The axial trough of the Red Sea represents a young tectonic depression. It starts south of 24° N lat. as a narrow, deep-water (1 to 2.2 km) fissure, gradually widening to the south and attaining 50—60 km in width. The axial trough in the southern part of the Red Sea gradually becomes narrower and tapers out even before the Bab-el-Mandeb Strait. Its northern terminal is less distinct. North of 24° N lat. the axial trough smoothens out and loses its morphological distinctiveness, although individual deep water basins (Gypseous, Kebrit, Oceanograph) are also noticeable here as well as along its continuation. It is also assumed that these might be traced up to the southern projection of Aquaba Bay in the Red Sea.

The complex constitution of the axial trough not only in transverse, but also in longitudinal sections was recently confirmed by researchers of the Oceanology Institute of the Academy of Sciences, USSR, namely, A.S. Monin, L.P. Zonenshain, O.G. Sorokhtin and others. Characteristic (92) features of the axial trough of the Red Sea, such as its high seismicity and heat flow, are well known. High amplitudes (up to 955.2 m A/m) of linear magnetic anomalies and the associated zones of positive (up to $+150 \cdot 10^{-5}$ m/sec^2) Bouguer gravity anomalies are its inherent features. Deep-water basins with hot brines and metal bearing sediments are confined to the axial trough and serve as the unique laboratory for contemporary ore genesis.

The bottom of the axial trough is highly dissected. The thickness of recent sediments is almost half of the main trough. They are more frequently confined to small grabens or 'pockets'. Traces of recent volcanic activities have been established from photography and dredging of the bottom and also investigations using a manned apparatus. The sedimentary rocks of the axial trough are underlain by rocks having a velocity of 7.08 ± 0.21 km/sec, which is characteristic of oceanic type crusts. Basalt samples dredged from the bottom are petrochemically similar to tholeiites of the mid-oceanic ridges of the Indian and Atlantic oceans. Investigations by the Oceanologic Institute of the Academy of Sciences, USSR established that the most recent volcanism in the axial trough could be localised in the narrow extrusive zone having a width not exceeding 0.5—1 km.

Thus, the combination of enumerated indicators of the Red Sea axial trough helps us to consider that its continental crust has been significantly changed, fragmented and replaced by a newly formed oceanic-type crust. The majority of researchers hold this opinion. However, views on the mechanism of the process vary. The problem on the amplitude of opening up of the Red Sea and the nature of the crust in the shelf stages of the main trough still remain controversial.

The absence of linear magnetic anomalies in shelf zones and the magnitude of longitudinal wave velocities of about 5.86 km/sec, on average, attest to the continental nature of the crust in the main Red Sea trough. Higher velocities, in the range of 6.6—6.7 km/sec, have also been established, which are more characteristic for oceanic-type crusts. Such contradictory values of longitudinal wave velocities in the crust of the Red Sea shelf zone reflect its specific character. This has been caused by the 'granitic' layer having been subjected to shear, fragmentation and intrusion of basalt dykes in the process of destruction of the continental crust in the early stages (Oligocene-Miocene) of the Red Sea basin formation. As a result, its thickness reduced sharply and the composition changed. All these apparently led to change in its physical parameters, closer towards values of the 'basaltic' layer. Data provided by D. Lowell favour this concept. Accordingly, shelf zone sediments in the southern part of the Red Sea were highly disturbed and tilted in the process of its shearing and thinning (without total disruption of its continuity). However, this also indicates injection of Miocene tholeiite basalts in the coastal zone of north-eastern Sudan as well as on the Jebel-Taif region in the Saudi Arabian coast.

(93) D. Ross and J. Schiller studied the southern part of the Red Sea in 1977, and in particular the unique area where a number of researchers had identified linear magnetic anomalies in the shelf. They concluded that the magnetic, seismic, structural and petrologic data do not confirm the presence of an oceanic crust under the entire Red Sea basins.

Thus, the main features in the constitution of the Red Sea rift are: 1) zoning, revealed in the symmetry of coastal escarpments, shelf zones and structural compounds of the axial trough; 2) segmentation revealed in the presence of three parts: northern (north of lat. 25°), central (between lat. 25° to 19° N) and southern (to the south of lat. 19°). The northern and southern segments are characterised by distinct linearity with a consistent and average north-east strike and relatively simple constitution of the rift components. The middle segment is characterised by more complex constitution and change of strike to submeridional, distinctly manifested along the western bank in the segment between lat. 19°–21° and lat. 23°–24°. The axial trough is highly dissected at the bottom in the central segment and grades into a series of separate deep basins. Almost all deep-water basins with hot brines and metal-bearing sediments (see Figs. 6, 16) are situated here. The oceanograph basin situated in the northern segment (lat. 26°40′) is an exception, but it is situated at a place where the axial trough undergoes weak, as though 'rudimentary', submeridional deviation.

FAULT DISLOCATIONS AND THEIR ROLE IN THE FORMATION OF THE RED SEA RIFT ZONE

A system of faults with north-west, north-east, meridional and sublatitudinal strikes are distinctly manifested in the structural set-up of the Red Sea rift zone. However, investigations have shown that submeridional and north-west-striking faults predominate and have determined the configuration of the main structural components, namely, the axial trough and the main basin, and their bordering marginal faults and uplifts. Linearity and parallelism of structures are characteristic for these elements which indicate their genetic commonality.

The leading role played by submeridional and north-west-striking faults in the constitution of the Red Sea rift zone explains the nature of the transverse bending of the rift. The Red Sea Rift distinctly deviates from its general north-west strike at lat. 18° and 23° N along the western coast. The coast line at these places forms the bend. Geomorphologically, depressional basins, deep cutting river basins occurring extensively in the continents in a submeridional strike, are associated with such bendings.

Geologic-geophysical investigations in the western framework of the Red Sea show that the bends in the rift are associated with submeridional faults, the largest of which are the Barak, Diib and Nile faults. Submerid-
(94) ional faults play the role of large structural components of deep-seated emplacement and had a prolonged period of development. They controlled the location of the multiphase alkaline ring intrusives that formed during the late Precambrian to Mesozoic and continued even up to the Cenozoic.

One such fault is the marginal one of the Red Sea basin, bordering it by a submeridional fragment. The fault is traceable along the rift escarpment of the Red Sea hills of Sudan from lat. 22° to 18° N for a length of more than 500 km. E.N. Isaev established that the fault is demarcated in geophysical fields in its contact zone with the Red Sea hills. An intensive gravitational stage is associated with it, reflecting changes in the deep-seated structures of the crust and the rise of the Mohorovičić surface (see Fig. 22). A gravitational maximum is confined to the fault at its place of transition to the continent [11]. The fault, bordering the Red Sea rift, appears to taper out into the continent accompanied by its deep-seated riftogenic reconstruction (see Figs. 20, 21). Farther south, the border fault transforms into the Barak fault and interruptedly continues up to Lake Tana in Ethiopia.

The other meridional structure of the Red Sea rift zone is the Diib fault (see Fig. 8). It has a strictly meridional strike and extends for not less than 800 km from Egypt in the north to Ethiopia in the south. The fault is represented by a system of dislocations which may be grouped into a zone having a width of up to 30 km and intersecting the Red Sea hills approximately submeridionally along the lat. 36° E. In the north, the fault begins with the bend of the Red Sea graben by a depression filled with Neogene-Quaternary sediments. Farther south, the fault zone is traced by a system of submeridional grabens filled with 'Nubian sandstones' of Cretaceous age. The grabens are, on average, 5 to 15 km wide and a few to several tens of kilometres long. The largest graben, situated in the northern part of the Red Sea hills, is up to 20 km wide and extends for 80–100 km. Smaller prefault grabens can be traced along the fault for more than 100 km. Farther south, in the central part of the Red Sea hills, the Diib fault zone is manifested in the form of submeridional grabens filled with conglomerates of indistinct origin (possibly late Precambrian molasse) and also manifested as a series of faults having the same strike.

According to E.N. Isaev and M.A. Ieda, the Diib fault zone, south of 17° N lat., is manifested by a system of grabens revealed by geophysical investigations. A series of meridional minima has been traced here in the gravitational field between lat. 17° N and 15° N. Considering the available geological data, these anomalies are interpreted as a system of graben-form structures filled with volcanogenic-sedimentary rocks of lower density. Analysis of gravimetric data shows that the depth of burial amounts to 2—2.5 km. The Gash graben is the largest among these structures, extending 80—100 km with a width of 10—15 km. It is asymmetric in the northern part and represented by gentle western and steep eastern boundaries. South of lat. 15°40' N the Gash graben widens, its boundaries become low-dipping, the burial depth decreases and the structure flattens out around Kassala city. Possibly, the graben totally withers out further south. However, considering the interrupted character of the Diib zone, the further continuation of this zone seems to be more possible in the southern direction along the boundary of the Ethiopian basalt plateau, where faults of a similar nature were earlier noticed by P. Moro and recently reconfirmed by J.R. Vail.

(95)

The meridional Diib zone of faults is one of the main features of contemporary structure in the Red Sea hills. It is deep-seated and underwent a prolonged period of development. The faults of this zone control the emplacement of late Precambrian linear granitic intrusions of 520 million years age (K-Ar method) and of the Mesozoic post-Nubian volcanism (trachytes, phonolites, lava-breccia), manifested in the form of volcanic plugs, necks and pipes. Besides, grabens traced along the zone localise relicts of Neogene-Quaternary basalt and rhyolite sheets, occurring on the 'Nubian sandstones'. The configuration of the relicts is also elongated in a submeridional direction.

Spatial distribution of ring intrusions of the Red Sea hills shows that they are controlled by the Diib fault and to a lesser extent by the Barak fault. At present, about 20 ring intrusives and stocks have been established in the Diib fault zone. The intrusives have been grouped into submeridional chains in individual segments and their relationship with faults evoke no doubt. Besides the detected massifs, numerous isometric magnetic anomalies, possibly reflecting ring intrusives which are not exposed by weathering, are localised in the fault zone.

That the meridional Diib and Barak faults underwent prolonged development is confirmed by the multiphase and prolonged formation of associated linear and ring intrusives together with alkali-basalt volcanism. Radiometric age data (K-Ar method) obtained by the author for the Salala,

Ankur and Umm-Shibrik ring intrusives indicate a wide age range for the formation of these structures. Thus, formation of the Salala ring complex took place during a 570 to 250 million year interval, Ankur—250 to 150 million years and Umm-Shibrik—180 to 140 million years. Most recent magmatic manifestations in the Diib fault zone are the basalt flows (120 million years) and alkali trachytes (85 million years) and in the Barak fault zone—basalt flows and alkali-granite intrusives (50 million years). Available information shows that manifestations of alkali-gabbroid magmatism are extended in time and belong to late Precambrian (Vendian), Palaeozoic, Mesozoic and Cenozoic. All these facts attest to the prolonged tectonic activity and magmatic permeability of meridional faults in the process of their evolution.

(96) The meridional Diib and Barak faults are the major structural elements of deep-seated occurrence and prolonged development. The magma-controlling significance of these faults, similarity in strike with East-African rift systems (rift links of Mobutu-Sese-Seko and Rudolf lakes) and their location in continuation of the latter prompt us to regard the Diib, Barak and other faults as the northern continuation of the East-African rift system. This analogy becomes particularly convincing if we consider that the Diib fault zone essentially represents a system of small grabens and horsts, occurring like 'miniature' rifts. O.P. Moro also noted the existence of the connecting link faults to the north from Lake Rudolf along the western escarpment of the basalt plateau in Ethiopia. Notably, V.G. Kaz'min later discovered ring intrusives in this region.

Major meridional faults were conduits for magma and preceded Cenozoic rift formations in the course of their development. Characteristically, the ring intrusives in the Red Sea zone are grouped in linearly elongated zones (belts) striking north-west. It has been noted that these zones with a width of 50 to 150 km border the rift depression and extend parallel to the latter for 50—100 km from the coastline. The alkali magma belts (zones) display genetic commonality with the Red Sea rift and are regarded as a unique form of the prerift activisation stage—activisation furrows dividing the platform into banded and to a varied extent activised blocks with a general north-west strike. We should note the coincidence in the configuration of belts with the structural set-up of the Red Sea rift. All these indicate the interrelationship of the Red Sea rift structural set-up with the adjacent faults and alkali magmatism zone.

A system of faults with north-west, north-east, meridional and latitudinal strikes is distinctly manifested in the structural plan of north-east Africa. These faults control the localisation of alkaline ring intrusives. They may be traced at places by chains of ring intrusives. Consistency of the strikes, their prolonged period of development and magma-controlling role enable us to see manifestations of a regmatic fault network in them. The existence of such a fault system is acknowledged by many researchers, e.g., S.S. Shultz, V.E. Khain, A.F. Grachev and others.

Analysis of the quantitative correlation between the orientation and distribution of faults in the Red Sea hills rift formation zone enable us to conclude that both the diagonal and orthogonal system of regmatic fault networks are equally manifested. Faults with submeridional and northwest strike predominate, signifying thereby that predominance of one direction took place among the orthogonal or diagonal systems.

As has been shown for the Red Sea the rifts used the framework of these faults in the course of their development in the Baikal and other rift zones. In the case of the Red Sea rift, where submeridional and north-westerly strikes predominate, the rift 'adjusted' itself to such as if it 'drifted' (97) in the said directions. This situation should be taken into account in determining the mechanism and movement kinematics of the Red Sea rift.

In the Red Sea rift formation zone the regmatic network existed even at the end of the Precambrian (Vendian). Faults of this network controlled the alkaline volcano-plutonic annular strictures and were actively developed during the Palaeozoic, Mesozoic, and Cenozoic [30]. Some authors hold that the regmatic network of faults had formed much earlier, possibly in the Archean. In the opinion of V.E. Khain, the characteristic system of narrow linear green-schist-bearing troughs dividing the Archean basement of the Aldan shield and South African massif may be considered as the manifestation of the regmatic network of faults.

The diagonal and orthogonal system of faults in the Red Sea rift zone is unevenly manifested both in time and space. The faults of one or the other trends were activated in different stages of development. Submeridional faults, on the whole, actively developed in the Palaeozoic-Mesozoic, although individual faults of a particular trend remained active also in the Cenozoic. In Saudi Arabia, for example, they controlled the emplacement of Palaeocene-Quaternary volcanism, and in Egypt, formation of the Nile River valley. In Jordan and southern Syria, north-west-striking faults predominated in the formation of the Druz Neogene-Quaternary volcanic trough. In Sudan, the faults of this strike control the El-Arab rifts of the White and Blue Nile, which are filled with thick (above 5000 m) Mesozoic-Cenozoic sedimentary strata. A combination of submeridional and north-western structural trends is distinctly manifested in the strike of the Sudanese rifts as well as in the Red Sea rift zone.

According to the author, by acknowledging the significant role of the regmatic network of faults in the structure of rifts, the areas of branching of individual rift units become more understandable. There are numerous cases in the African-Arabian rift belt where the rift structure on the whole or its individual faults are displaced towards the left or right of the strike. V.G. Kaz'min and others explain such bends in trend of rifts in plan as due to the presence of transformed faults. E.E. Milanovskii considers that en echelon linking of rift structures or their branches took place in this case.

The author considers that in the presence of a network of faults preceding the rift formation process, the latter may, in some segments, have used the submeridional faults, and, further, become combined in their strike with north-eastern or north-western-striking faults, and rarely with the submeridional ones, thus deviating from the initial strike. Further, the rift may again have followed the submeridional strike. Such an approach to the formation of rift structures does not exclude a transformed nature (in a broader sense) of transverse and diagonal faults in plan and their prerift occurrence. However, adequate data for unambiguous inclusion of these faults into the transformed ones are not available at present. Often, (98) their kinematic nature has not yet been established and, in the opinion of the author, the fact that in a number of cases rift faults intersect transverse faults and do not significantly displace them, prevents the identification of the latter from the transformed faults of mid-oceanic rifts.

It may be noted in this regard that the northern continuation of the western branch of the east African rifts is one such place where the occurrence of a transformed fault is assumed. In actual practice, the western rift is shown as disrupted due north of Lake Mobutu-Sese-Seko in the au courant geological maps and tectonic schemes. It is presumed that the branch ends by butting against the Asva fault, which is assumed to have the role of a transformed fault. A. Bateman presumes that individual negative anomalies, traceable from northern Uganda, may indicate their possible continuation in Sudan. The author has already emphasised that the Nile valley in northern Egypt is tectonically controlled and is confined to submeridional, north-eastern and north-western faults and its geniculate form corresponds to the points of intersection of these joints. The possible relationship of faults, controlling the confinement of the valley with the western rift, has already been observed.

More data on the existence of rift structures in central and southern Sudan have appeared in recent years. As a result, three branches of the rift (Malakal-Kosti, Bara and Juba-Abu) are distinguished in Sudan according to the data of El-Raba, El-Shafi, R.B. Salama, P.M. Bermingham and others. They are filled with about 4 km thick Neogene-Quaternary continental sediments of the Umm-Ruvaba series. It may be considered on the basis of these data that the East African rift system continues into southern Sudan. A combination of submeridional and north-western structural directions, which are characteristic of the Red Sea rift-genesis zone, is distinctly manifested along the strike direction of the Sudanese rifts (grabens).

Continuation of the western rift to the north is further confirmed as a result of associating major meridional faults controlling the distribution of

ring intrusives of alkaline composition and from new data on the constitution of the Upper Nile Basin. A major submeridional fault represented in
Egypt by a system of faults in the Nile valley is detected in the northern
continuation of the western rift. Linearly elongated outcrops of granites of
600 million years of age coincide with this fault in the Aswan region, indicating thereby its earlier emplacement and similarity with the Diib fault in
this respect. Farther south, the fault is traced by the submeridional Vadi-
Halfa valley. Outcrops of ring intrusives coincide with it in the meandering
regions of the Nile north of Khartoum city in central Sudan. Farther south,
the fault is traced by a submeridional fault, named by the author the Nile
fault.

The fact that the upper Nile basin has a graben-type environment
and is filled with continual deposits of the same age as that found in
the Red Sea basin, brings the Nile fault zone closer to rift structures.
(99) The possible structural relationship of the Upper Nile basin (rift) with the
western rift is also emphasised by the association of alkali ring intrusives
to the branching zone in southern Sudan.

Along with the submeridional and north-west-striking faults, a significant role in the constitution of the rift must be assigned to faults having
a north-eastern strike. Many researchers have observed the presence of
faults of the same direction in the axial trough. R. Girdler was the first to
propose that transverse faults of a north-eastern strike are traceable from
the water areas (marine expanse) to the continent. In his opinion, the
transverse fault of the axial trough, passing into the land mass in Tokar
region south-east of Port Sudan, is one such fault. Based on bathymetric
as well as magnetometric and seismic data transverse faults were observed later by many researchers in the structure of the axial trough of
the Red Sea rift.

D. Philips established on the basis of his study of the magnetic field of
the axial trough in the region between lat. 20° and 22° N that expansion
of such faults took place in the 60° NE direction and displacement along
the transformed fault with the same strike passing through the Atlantis
II basin. It has also been established that the majority of epicentres of
earthquakes are confined to the central part of the axial trough between
lat. 19° and 21° N, where major transformed faults might be identified.
Assessment of the focal mechanism for earthquakes in the Suakin basin
(19°8′ N lat., 38°8′ E long.) and in the region of the submerged Ramad
peak (17°2′ N lat., 40°6′ E long.) showed that they conform to thrusts
with laevo displacement along north-east-striking planes, corresponding
to NE 43° and 49°.

The structural plan of the Red Sea hills of Sudan, revealed from the
subdivision of the territory into north-east-striking blocks that are separated by gravitational steps and linear zones of differentiated magnetic

simulation, continues into the waters of the Red Sea. Faults bordering these blocks in the continent are revealed in the bathymetry of the floor, both in shelf zones of the main trough as well as of the axial trough. Transverse (transformed) faults, established by magnetic field and bathymetry, are situated along their continuation in the axial trough of the Red Sea. It may be noted that deep-water basins such as the Atlantis, Discovery, Waldivia and others are located, in the majority of cases, along north-eastern-striking faults of the continent in the axial trough of the Red Sea.

North-east-striking faults in the western boundary of the Red Sea, formed in the orogenic stage (late Riphean-Vendian) of the late Proterozoic mobile belt, maintained their activity in much later epochs. They controlled the distribution of ring intrusives and 'layered' gabbro intrusives over a prolonged period of development, namely in the Vendian, Cambrian, Ordovician, Silurian, Devonian, Jurassic, Cretaceous and Palaeocene.

(100) Nine major band-like (linear) anomalies of magnetic fields, coinciding with the NE-striking faults, were distinguished based on the geologic-geophysical data on Sudan, Egypt and maps of magnetic fields of the central part of the Red Sea and its boundary (Project 'Atlantis II'). They could be established from geophysical data (gravitational steps), displacement of geologic structures, linearly controlled ring intrusives, geomorphological and other indications (from north to south). The subdivisions are as follows: 1) Southern Egypt, traceable into the deep-water basin of the Kebrit axial trough; 2) Nugrus—into the Venus and Waldivya basins; 3) Halaib—into the Nereus basin; 4) Dungunab (Sofai)—into the Tetus basin; 5) Mohamedkol (Amarar)—into the Atlantis and Discovery basins; 6) Port Sudan—into the Erba basin; 7) Sinkat—into the Port Sudan basin; 8) Derudeb—into the Suakin basin; 9) Karor—traceable into the flexure of the axial trough at 18° N lat. (see Fig. 19).

Such excellent similarity in the location of deep-water basins of the Red Sea axial trough, its transverse faults and faults in the continent cannot be of a random nature. It may possibly be explained by the fact that localisation of basins is predetermined by the structural framework of faults which preceded the formation of the main trough of the Red Sea. A network of north-eastern and submeridional faults preceded the formation of the Red Sea basin in the prerift stage. The north-eastern faults under review are the components of this network in which, it appears, the rift, so to speak, became fixed. In such cases the north-eastern faults played the role of shear planes along which transverse displacement of the Red Sea rift and its structural components, namely the axial trough and shelf stages, took place during subsequent reactivisation.

It is interesting to note that the deep-water basins of the axial trough are not irregularly distributed. It may be presumed from a certain amount of conceptualisation that the network of faults has a geometric regularity, i.e., the distance between the faults is equal. The correctness of such an assumption is confirmed by the fact that north-eastern faults are situated at approximately equal distances, in any case, between the main faults. This is particularly evident in faults south of the Arabian Desert in Egypt. A similar distance in location of deep-water basins of the axial trough of the Red Sea and of the north-eastern faults in the continent helped in identifying an 'advance' of the fault network in its formation and to conceive a commonality in their structural control. From such standpoints, the location of deep-water basins of the axial trough may be assumed to have been 'fixed' in the framework (network) of faults. The junction of intersection between the network of faults is the most possible location of deep-water basins. It is noteworthy that bends of the axial trough in such cases when they are not accompanied by deep-water basins (25°30′, 22°00′, and 19°00′), are also recorded in the 'advance' of the faults network and this fact indirectly indicates the possibility of revealing deep-water basins or their accompanying hydrothermal processes in such places.

(101) The above-noted interrelationship of north-eastern faults, having a prolonged period of development and deep-seated disposition in the continent, with the transverse faults of the axial trough and deep-water basins containing hot brine and metal-bearing sediments, may be considered as an indication of 'adaptability' of the rift-forming process and the ancient anisotropy of the basement ('frame') on which the Red Sea rift was formed. Such excellent similarity in the location of deep-water basins of the axial trough, its transverse faults and faults in the continent indicates that the structural-material complexes that are reflected in magnetic fields and the faults turned out to be more complementary to rift genesis. The endogenetic process in deep-water basins of the axial trough had, as it were, 'burnt out' the substrata to the maximum in strictly predetermined places, which once again denotes the dependence of rift formation upon the structural anisotropy of the basement and the discreteness of rift formation along a strike.

North-east striking faults of the Red Sea rift are conformable with the shear direction and displace the axial trough from the oceanic type crust, i.e., their transformational role is manifested here. These data show that the transformed faults follow the ancient Precambrian faults and also that their development and location in the structural plan of the rift zones have a predetermined character.

Investigations within the western boundary of the Red Sea (Egypt, Sudan) and also throughout the entire Red Sea rift zone show that latitudinal faults have great significance in the structure of the rift zone. It has been established that latitudinal faults intersect Precambrian metamorphic and intrusive rocks, and, as a rule, are distinctly revealed in the relief by river valleys filled with Holocene alluvial-proluvial deposits. They often form a characteristic structural-morphological set-up in combination with meridional faults, which has been termed the chessboard type by E.E. Milanovskii for the Kenyan rift. The latitudinal as well as meridional faults are of ancient origin. However, it is not possible to determine their exact age of development. They were formed mainly in the Cenozoic. Faults of this trend disrupt the Neogene sediments of the precoastal plains and also their contact with the Precambrian. The latitudinal faults have been the most active in the recent period among all the fault systems within the limits of the Red Sea. Intensive valley erosion has taken place along them in a direction transverse to the strike of the Red Sea rift. Their most recent activity has been confirmed by tracing them in contemporary river valleys. More major latitudinal faults in the western boundary of the Red Sea are the Mersa-Alyam Belts in Egypt (lat. 25° N), Mohamedkol and Hor-Shinab faults in north-eastern Sudan. The Hor-Shinab fault is 70 km long and the Mohamedkol fault is 90 km long. The Mersa-Alyam fault is approximately of the same length. All the above-mentioned faults are of the thrust type with dextral displacement and 1.5–2 km amplitude. Determination of vertical amplitude is difficult. Judging by the interrelationship of the height of the river embankments, it seems that possibly 02) the vertical displacement does not exceed a few hundred metres. The latitudinal faults along with the meridional ones form an orthogonal system, which plays an important role in the structural set-up of the Red Sea rift zone. It is important to note that faults of this strike direction are extensively developed in the axial trough of the Red Sea as well.

The foregoing shows that faults of north-west, north-east, meridional and latitudinal strikes played a significant role in the structure of the Red Sea rift zone. The cited data leave no doubt about the ancient age of the faults and predetermination of the strike and structural set-up of the Red Sea rift zone by them, which may be considered as a planetary network. At present, the planetary significance of network faults is not disputable. Its significant effect on the location of rift zones in plan is displayed well for the Baikal rift zone and East African rift belt. The significance of the planetary network faults in the structure of rift zones on the whole and particularly for the African-Arabian zone increased sharply after the revelation of cosmogeologic investigations. The fault maps for Afar and adjacent parts of Ethiopia, compiled by P. Kronberg and others from cosmic photographs, show that north-east and north-west are the predominating structural directions. Faults of the orthogonal system are weakly revealed.

Results deciphered from cosmic photographs over large territories, including not only the Red Sea rift zone, but also the African basin with

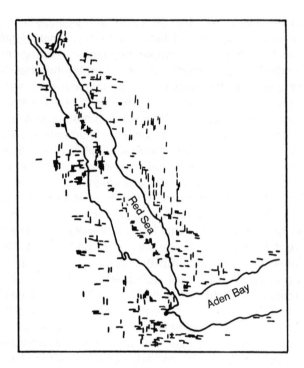

(103) Fig. 23. Orthogonal system of faults in the Red Sea rift zone (after M. Shoenfeld).

adjacent parts of the Ethiopian plateau and bordering areas of Aden Bay showed that the following systems of faults are distinctly manifested at these places. Orthogonal (Fig. 23) and two diagonal (Figs. 24, 25), one of which is exactly parallel to the general strike of the Red Sea (north-west, 330° and north-east, 45°), the other strike north-west, 315° and north-east, 55°. Bathymetric data of H. Becker on the axial trough of the Red Sea proved even more important. According to this concept, the higher isobath gradients of the axial trough reveal distinct linear escarpments following the faults. The escarpments are associated with faults constituting the same network (of faults), revealed from cosmic photographs within the Arabian and African framework of the Red Sea. Besides the general strike of the axial trough, the isobath gradients depict more or less sharp and consistent changes in direction and steepness of slopes, associated with most recent faults. Faults distinguished from the axial trough bathymetry have, in the majority of cases, a north-west strike, and are intersected by faults of a north-east direction. Latitudinal and meridional faults have also been distinguished. The existence of lateral and transverse faults in the axial trough structure of the Red Sea was recently confirmed by direct observations from the manned underground apparatus 'Paisis'.

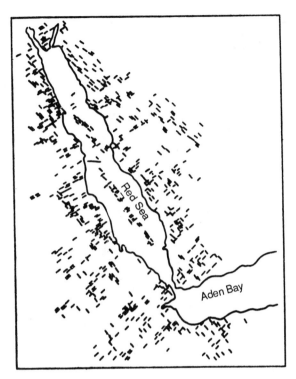

Fig. 24. Diagonal system of faults (north-west 315° and north-east 55°) in the Red Sea rift zone (after M. Shoenfeld).

While delineating the diagonal system of faults, it was observed that one of the directions strictly paralleled the general strike of the Red Sea basin. It is important to note that there exists an extended zone of aeromagnetic anomalies, consisting of narrow negative and positive anomalies with sharp gradients, and the same extend parallel to the basin along the Saudi Arabian coast for more than 1500 km and with a width of about 100 km. The magnetic anomalies are located in Precambrian complexes and have an intersecting relationship with the latter. The more intensive anomalies were caused by steeply dipping dykes of Miocene age (22 million years from K-Ar method), having reverse residual magnetisation. Other anomalies are caused by dykes with normal residual magnetisation and steeply dipping faults. The dykes vary in composition from dolerites to quartzmonzonites. Possibly, Oligocene (about 30 million years) tholeiite dykes of the Tikhama-Asir region are the oldest in the system. The dyke belt has been observed to be interrupted at the latitude of Jedda city at the place where transverse transformed faults of the Red Sea axial trough are exposed on the coast. Magnetically excited tholeiite dykes of

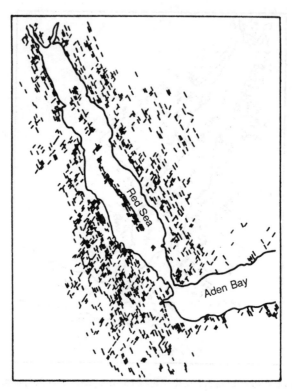

(104) Fig. 25. Diagonal system of faults (north-west 330° and north-east 45°) in the Red Sea rift zone (after M. Shoenfeld).

the same strike have been identified in the Eastern Egyptian desert on the African coast within the boundaries of the Red Sea. It has been presumed that the Arabian dyke complex is associated with a deep-seated fault, along which the Precambrian rocks of the Arabian block were downthrown in the process of formation of the riftogenic continental boundary.

(105) Precise parallelism of the dyke zone with the Red Sea basin undoubtedly denotes a genetic relationship in their formation. Existence of linear magnetic anomalies in the continental crust, genetically related with rift formation, implies the possibility of their (later) formation without drawing on the 'spreading' model of the tectonic plate concept.

Analogous linear but diversely oriented magnetic anomalies have been established in the Afar basin where the continental crust thinned and became considerably stretched but did not tear off. This problem will once again be dealt with in Chapter 5, while discussing the model of the Red Sea basin formation. It is to be noted here that the magnetic anomalies in the main trough of the Red Sea in its south-western part in the Dahlak archipelago region has possibly the same nature. Magnetic anomalies of 40 to 35 million years age were detected here by a few researchers as indicators of the oceanic crust. However, these anomalies are not extended and were not detected in other parts of the Red Sea basin. Therefore an attempt to consider on the above grounds that the crust of the main trough of the basin is oceanic and, moreover, formed by a 'spreading' model, would be fruitless. Moreover, seismic investigations of R.B. Girdler in this part of the Red Sea showed that a layer of 5.91 km/sec velocity lies under a sedimentary layer with 3.3 km/sec longitudinal wave velocity. The former is characteristic of crystalline rock, downthrown by faults, although to some extent compacted due to injection of basaltic material, and was detected in the course of drilling for petroleum in the form of sills and flows.

The author also obtained interesting data from a comparison of the Red Sea bathymetry with the Precambrian structures within its boundary. Thus, the sea bottom possesses a fairly complex relief at places within the shelf zone (depth below 1000 m) of the Red Sea where boundaries of the rift zone intersect the Precambrian structures at a steeper angle. Depressions and projections of the sea bottom and also the limitation of shelf and preshelf zones are governed by Precambrian structures in their orientation and possess a predominantly north-eastern strike like the latter. In contrast, the morphology of the shelf and preshelf zones is comparatively simple in those places where limitations of rift depressions and Precambrian structures intersect at a gentle angle, and their boundaries are predominantly linear with rare local underturns conforming to the orientation of the Precambrian structures.

Thus, establishment of faults of ancient (Precambrian) origin within the Red Sea boundaries and subsequent prolonged (periodic) activisation practically throughout the entire Phanerozoic, as manifested in their magma-controlling role, are confirmed by the decoded cosmophotographic and bathymetric data of the Red Sea bottom. Parallelism between data on faults and data on the Red Sea axial trough with the newly formed oceanic type crust helped to conceive a genetic relationship among them. This fact may be interpreted as an indication of (106) the formation and development of the main as well as axial trough under the effect of the ancient system of faults controlled by the basement of the Arabian-Nubian shield. It is important to emphasise that detection of the consistent network of faults sharply limits the possibility of rotation of the Arabian block around the pole located in the eastern Mediterranean Sea, as has been assumed by a number of researchers in explaining the kinetics and mechanism of development of the Red Sea and Western-Arabian (Levantin) rift zone.

In the light of establishing a singular system of faults within the continental framework and in the axial trough of the Red Sea certain questions arise: How was the constancy of the systems of faults in the continental and oceanic crust preserved (if the latter exists in the axial trough in 'pure' form) and why did the oceanic crust react with the continental crust in the same way for later tectonic stresses? It may be presumed that ancient faults penetrated into the newly formed oceanic crust due to the common stress fields caused by a deep-seated subcrustal process and the action of global forces, possibly associated with the earth's rotation, which were activised in the continental crust. However, it is difficult to explain from these positions why the newly formed oceanic crust, representing growth of basalt columns corresponding to the 'spreading' model and symmetrical with respect to some centre, must 'be split' along ancient Precambrian faults of the continental crust, which are discordant to these columns. And one more issue remains unclear: Why does the axial trough not only manifest faults conformable with its strike and transverse faults which are widely developed in oceanic rifts, but also with meridional and latitudinal faults? It is naturally assumed that structures of the lower parts of the continental lithosphere continue their 'residual' effect in the axial trough of the Red Sea and, as if, 'illuminated' in the newly formed oceanic type crust, and that in the given phase the axial trough does not yet represent itself to be a 'separated' and gaping fault, filled with basalt, for the entire depth in the lithosphere. The destruction of the continental crust was achieved in the mid-Aden and mid-oceanic ridges with the latter stage. It may be presumed in this regard that the oceanic type newly formed crust in the axial trough of the Red Sea did not completely lose its links with the continental lithosphere. The crust could have persisted in this case in a 'pseudo-oceanic' stage, when a thin intermediate layer possibly still existed in it that had been so greatly saturated with magmatic material from the mouth that to distinguish it from the typical oceanic crust is now difficult.

The aforesaid suggestions could lead to the conclusion of a deep-seated (lithospheric) level of emplacement of the planetary network of faults in the Red Sea rift zone. This is also denoted by such data as transition of major faults (lineaments) from the continent into the ocean (or the reverse) and inherited development of oceanic transformed faults from the ancient (Precambrian) faults in the continents. The conservative nature of the network of faults in the Red Sea rift zone is contradicted by (107) major displacements of the Arabian block with its anticlockwise rotation, as has been suggested by many researchers.

TIME OF FORMATION OF THE RED SEA BASIN

Voluminous literature has been published on the problems of origin of the Red Sea basin.

The formation of the basin was long treated from the arch model viewpoint of H. Cloos. The idea of separation in the context of the lithospheric plate tectonics concept has developed in the last 10–15 years and might be applicable here also. The basin has served as the object for innumerable geologic-geophysical investigations. At present, data are available in sufficiently summarised form in many publications in the USSR and abroad. It is to be noted that mainly the same factual data are presented in many publications but interpreted from different points of view. The majority of researchers are unanimous about the graben-form structure of the Red Sea depression and that the main stage of its formation took place during the Neogene-Quaternary period. However, the time of emplacement of the basin is debated because the possibilities for studying the earlier (prerift) stage of development of the basin were formerly limited, since the basin is mainly composed of Precambrian rocks. The Palaeozoic and almost the entire Mesozoic stage have been excluded from analysis, except in a few cases.

After stabilisation of the late Proterozoic fold belt of the Red Sea during the late Riphean-Vendian, the remaining Red Sea rift territory underwent extensive arched upliftment and transformed into the Arabian-Nubian shield stage of development. The Arabian-Nubian shield underwent consistent upliftment during the Phanerozoic stage of development, encompassing the Caledonian, Hercynian, Kimmeridgian and early Alpine (Late Cretaceous-Eocene) stages. Prerift activisation processes were extensively manifested in all the stages in its central part. They split the shield into rigid and mobile blocks, permeable for alkaline magmatism of the belt, i.e., activisation furrows, such as the Arabian in the east and the Sudanese in the west. Occasionally, the northern and southern periclinial terminals of the shield underwent local burial, when marine pericratonic basins, such as the Western Arabic (eastern Mediterranean) in the north and the East African (Indo-Chinese) in the south, transgressed into the shield and penetrated deep along its axis, but never became united. This refers to the coal deposits of Suez Bay and the Cretaceous-Palaeogene sediments in the northern part of the Red Sea. During the late Cretaceous, the marine basin penetrated into the central part of the shield up to the latitude of Jedda city where shallow-water marine sediments with fauna of Maestrichtian age could be observed [36]. Moreover, the increased thickness of the coal deposits in Suez Bay and its coincidence with the Devonian trough might denote that formation of individual segments of the Red Sea structural trends, particularly of the Suez trough, (108) took place at different times. However, extrapolation of the coal-bearing trough in the Red Sea is less probable and practically unfounded. The characteristics of development of the Red Sea rift become clearer in the light of E.E. Milanovskii's assumption regarding discrepancies of development of linear platform structures both in time as well as along the strike. The location of the deep-seated Marda fault and its prolonged development in continuation of the Red Sea basin in Ethiopia and Somali is remarkable from these viewpoints. The fault may be regarded as a major linear platform-based Indian-Mediterranean lineament, extending for more than 3000 km from the Mediterranean Sea through Suez Bay, Red Sea, Afar basin and Marda fault up to the continental slope of the Indian Ocean. The lineament had prolonged development; its formation in the late Precambrian is established by its association with the Precambrian dykes and fissures in the Sinai Peninsula as well as in the Red Sea hills of Sudan, magma-controlling role of submeridional faults of the Red Sea coast and Sudan in the late Riphean-Vendian-lower Paleozoic, late Precambrian fault formation in the zone of Afar basin framework in Ethiopia and relationship with the Marda fault and late Precambrian granite dykes in Somali.

Thus, the Suez graben, being a part of the Red Sea structural direction, became activised in the Devonian and Carboniferous along the ancient Indo-Mediterranean lineament. As regards the Cretaceous and Palaeogene rocks preserved in Kuseir and Safaga grabens in the Red Sea coast, the comparative characteristics of their more interiorly located regions of the platform (Nile Valley, Harga and Dahla oasis) showed that they possess common lithofacies and similar thickness (Fig. 26), displaying thereby the commonality of tectonic events in the Cretaceous-Palaeogene stage. The reliability of such comparison is based on detailed lithofacies and zonal microfaunistic classification of Cretaceous and Palaeogene deposits conducted by V.A. Krasheninnikov. The thickness of the Cretaceous and Palaeogene deposits of all three regions is comparable and the variations insignificant. Some increase in thickness towards the Red Sea (Kuseir) should not be regarded as an indication for the existence of a relative depression in the territory of the latter, inasmuch as the thickness of the Cretaceous and Palaeogene sediments diminishes again in the eastern part of Kuseir. The existence of a relative uplift in this region, in turn, cannot be extended to the Red Sea territory. In fact, the Nile River valley shows similar sharp changes in thickness.

The absence of any fundamental difference in the constitution of the Cretaceous-Palaeogene sediments convinced the author that all three regions belonged to a single sedimentary basin. It may be assumed that the Arabian desert territory in Egypt was a landmark during sedimen-
(109) tation but absence of lithologic changes in rocks in the direction of the assumed landmass, both from the western (Nile valley) and also the eastern (Kuseir, Safaga) side of the uplift, totally excludes the possibility. For this reason the grabens in the Red Sea region (Kuseir, Safaga) cannot be regarded as the preliminary sedimentary trough. Consequently, the

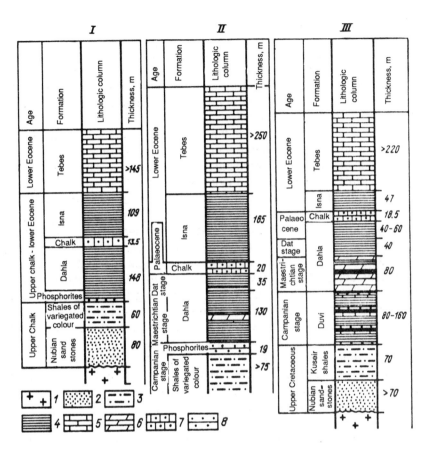

(109) Fig. 26.Correlation of sections of Cretaceous and Palaeogene sedimentary deposits in the Red Sea: Kuseir (III), Nile valley (II) and Harga oasis (I).

1—Precambrian basement; 2—'Nubian sandstones'; 3—clays, sandstones and marls of variegated colour; 4—clays; 5—organogenic and pelitomorphic limestones; 6—marls and chalk-type carbonates; 7—chalk-type carbonates; 8—phosphorite-bearing deposits (limestone, marl, phosphorite, silica).

tectonic activity (gradient of movement) of the Red Sea territory in the Cretaceous-Palaeogene stage of development did not essentially differ from individual parts of the platform, and is generally commensurable with it. The observed changes in the constitution of the Cretaceous-Palaeogene sediments were inherited by synchronous formations of the surrounding parts of the platform and do not signify individuality in the structural development of the Red Sea. Decrease in thickness, curtailment of stratigraphic volume and change of marine environments in the southern direction by shallow-water coastal marine sediments show that

(110) the marine basin had been bounded by the northern part of the Red Sea terrritory.

Thus, to speak about the formation of the rift in the Cretaceous and Palaeogene (up to Oligocene) would be without basis, insofar as sediments of this age are absent in its major parts: Their presence in the northern part of the rift (Kuseir and Safaga regions) is not related to the development of the Red Sea rift.

The conclusions reached on the developmental character of the Red Sea region in the Cretaceous-Eocene stage has not only theoretical significance in the aspect of confirming the formation time of the Red Sea rift, but also practical significance, insofar as industrial phosphorite deposits of the Nile basin are associated with the Campanian-Maestrichtian sediments. First of all, this fact helped in reorienting the prospecting method for phosphorite in the Nile phosphorite-bearing sediments. In distinction to prevailing concepts on the dependence of phosphate occurrence of the Nile basin and configuration of its ore-bearing regions on the suggested upliftment, adjacent to the development of the Red Sea rift, concepts were proposed regarding local and regional control of phosphorite occurrence. The first concept was arrived at from the development of synsedimentary uplifts and furrows, and the second from the location of the region in the marginal structural-facies zone of the epicontinental basin.

Subsidence of the Red Sea rift basin together with the Suez graben started in the Oligocene (?)—early Miocene (Fig. 27). It is associated with accumulation of variegated coloured quartzose sand formation deposited uncomformably on underlying rocks of different ages. The rocks in the Suez graben are predominantly well-sorted quartzose sands. Based on the study of sections exposed in the Red Sea coast of Egypt (Umm-Rusas) and north-east Sudan, the author concluded that the precoastal areas of the basins were composed of coarser grained rocks. They are represented in Egypt by thin-bedded carbonate-bearing coarse sandstone (60 to 80 M) and in Sudan by the Khammamit formation. The latter comprises reddish-brown coarse-grained quartzose sandstones with rare thin lenses of conglomerates having well-rounded quartz pebbles.

The formation of the Red Sea basin was accompanied by the short-lived but relatively faster growth of the upliftment of its surrounding areas (shoulder). The fact that coarse fragmentary composition is not characteristic for basal horizons filling the near coastal parts of the depression and situated in direct proximity of sources of weathering indicate that the formations of the basin took place under conditions of relatively quiet burial and absence of riftogenic upliftment around its boundary took place later in the Pliocene-Quaternary period. The Arabian-Nubian shield on the whole underwent moderate upliftment at the time of formation of the basin. It is important to keep this aspect in view insofar as the opinion

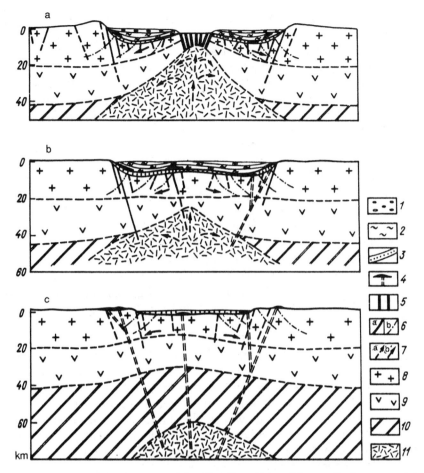

(111) Fig. 27. Scheme of development of the Red Sea rift: a—Pliocene-Quaternary period, b—middle-late Miocene, c—Oligocene-early Miocene.

1—terrigenous carbonate deposits; 2—evaporite series; 3—quartz sands, sandstones, conglomerate lenses; 4—basalt sheet and their leading canals; 5—cluster of basalt dykes in newly formed oceanic-type crust in the axial trough; 6—rift faults (a) and radiating fault dislocations in the brittle upper part of the granite-metamorphic layer (b); 7—tectonic stresses (a—in mantle diapir, b—in lower part of sialic crust); 8—granite-metamorphic rock layers; 9—'basalt' layer; 10—upper mantle; 11—anomalous mantle (roof of mantle diapir).

that formation of the Red Sea basin took place in a major upliftment is still put forward in literature. It is to be noted that a comparison of the (111) Neogene sediments, studied by the author in the Red Sea coast of Egypt between lat. 24° N and 25° N, in the Halaib region of the Sudan 'Nubian' sandstone-bearing regions in the Nile Valley of Egypt and the White Nile region south of Khartoum city in Sudan, shows that the Neogene basal

horizons and 'Nubian' sandstones are close to each other in quantity of detrital material as well as in their volumes. Increased coarseness of detrital material and some increase in their quantity from the south to the north have been observed in both the strata. These data may be interpreted as an indication of the fact that the contrast of relief of the Arabian-Nubian shield on the whole was apparently greater than during the accumulation
(112) of the 'Nubian' sandstones at the time of formation of the Red Sea basin in the Oligocene(?)—early Miocene.

The Khammamit formation is not characterised from faunal content. R. Carella, N. Skarpa and subsequently A. Bateman considered that this formation might be of Eocene to lower Miocene in age. However, in lithological composition, facies position of rocks and position in the section, the Hammamit formation is distinctly comparable with basal sandstones and gritstones of the lower Miocene in the Egyptian Red Sea coast, occurring in the coast of Suez Bay in thin-bedded quartzose sands of Oligocene age. The Hammamit formation and its analogues in Egypt form the Neogene basal red-coloured continental molasse sediments, corresponding to the early stages of formation of the Red Sea graben and formation of short flat upliftments bordering it. Let us deal with the stratigraphic position and role of the Mukkavar formation in the structure of the basin.

In the Maghersum borehole, the formation has been detected to underlie the Hammamit formation. The formation comprises dark grey, rarely reddish-brown siltstone-clays, interbedded with red and grey, fine- to medium-grained quartzose sandstones. Marls and limestones have limited significance in its composition. The nature of its contact with the underlying basalt tuff-breccia as well as the age of the latter are not clear. R. Carella and N. Skarpa suggest that the tuff-breccia might even belong to the Precambrian, i.e., the formation overlies the basement rocks. They also suggest that this formation unconformably underlies the Hammamit formation. The sediments of the Mukkavar formation contain mollusc shells foraminifera and ostracods which are identified with difficulty. Based on the presence of ostracods, these authors consider the formation to be of upper Cretaceous to Palaeogene in age and categorise the same with the Maestrichtian marine marls of the Jedda region in Saudi Arabia. However, it should be noted that identification of these sediments as well as determination of their age cannot be regarded as unambiguous, inasmuch as palaeontologists observed later that the ostracod complex is poorly preserved and its forms cannot be accepted as index fossils.

Thus, distinction of the Mukkavar formation and, more so, establishment of its age need still further confirmation, since these questions concern one of the fundamental problems of the Red Sea basin geology, namely its period of formation. It should be noted in this regard that a different interpretation is possible based on the age of the Mukkavar formation, according to which the latter represents marine shallow-water or fresh-water facies of the Miocene basal sediments (Hammamit formation). It is felt that the latter interpretation conforms better with available geological data. In this regard it is noteworthy that the Mukkavar formation has been found at places distant from the eroded upliftment of the Red Sea hills in the vicinity of which the Hammamit formation occurs directly upon the basement rocks. The concept of the Mukkavar formation as a facies or as a part of the Hammamit formation conforms with the composition of the basal layers of the lower Miocene in the Red Sea coast of Egypt where they are represented both by detrital red-coloured facies as well as by marine carbonate with calcareous sandstone horizons, also containing mollusc shells, coral and echinoid fragments.

The next stage in the formation of the Red Sea basin started in the middle Miocene and attained its maximum growth in the late Miocene. It is associated with accumulation of the thick evaporite beds of the Maghersum formation. Lacustrine-marine environment was consistent during this period for an enormous territory, which is displayed by extensive development of this formation along the entire Red Sea coast. The significant thickness (up to 1500 m) of the Maghersum evaporite formation indicates a sharp increase in the amplitude of depression of the basin. At this stage the Red Sea basin had already attained the dimensions close to its contemporary size, insofar as the evaporite beds are widespread in its entire territory excluding the southern part of the axial trough. Intensive downbuckling was accompanied by fissuring along faults and shearing of the downthrown and thinned continental crust, differential shift of its blocks which favoured eruption (extrusion) of basalts within the basin and its contiguous territories.

The end of mid-Miocene was marked by the accumulation of corallite limestones in the Abu-Imam formation. These formations are persistent over considerable areas and show relative tectonic calmness, causing homogenisation of the palaeogeographic set-up. Moreover, the relationship of Abu-Imam limestones with the underlying evaporite deposits is unconformable, which indicates manifestation of considerable tectonic movements before their accumulation. As a result of these movements, the evaporite strata of the Maghersum formation is totally missing in individual parts of the Hebel-Hammamit geological section, and limestones with traces of hiatus directly overlie the sandstones and conglomerates of the Hammamit formation. It is important to note that here the Maghersum formation does not contain gypsum and is represented by thin-bedded red clays and sandstones in the section. A more terrigenous composition of the Maghersum formation and its occasional absence from the Hammamit horst-anticline section exhibit the differentiated character of tectonic movements in the adjoining upliftment zones of the Red Sea hills and the Red Sea. Sediment accumulation took place under conditions of uneven upliftment or sinking of individual blocks represented by horst-anticlines and graben synclines in contemporary structure. The mentioned Hammamit horst-anticline was one such actively rising block at the end of the Mid-Miocene stage.

The succeeding stage of development of the Red Sea basin was marked by revival of sinking and accumulation of thick-bedded evaporite beds constituting the Dungunal formation. On comparison with the evaporite series of the regions farther north of the Red Sea coast, the latter corresponds more appropriately to the middle-upper Miocene.

(114)   The structure of the Red Sea basin became complicated in the Pliocene. The main stages of newer upliftments within the Red Sea framework took place, namely during this period. The movements were of the block upliftment type. The formation of Abu-Shagara and Hammamit horst-anticlines and the Dungunab Bay graben-syncline separating them on the Red Sea coast of north-eastern Sudan was concluded as a result of multidirectional displacement of blocks along their boundary faults with a predominant north-west strike.

The succeeding Quaternary period of development of the basin is associated with eustatic fluctuations in sea levels. The erosional depositional surface of continental and marine genesis formed during this stage on the Red Sea coast. The older erosional level is represented by the horizon of accumulated boulder-pebble conglomerates and shingles in the form of vast alluvial-proluvial trails in the foothills of the then fast-growing Red Sea hills uplift. Later rejuvenation, possibly towards the end of the mid-Pleistocene, led to the upliftment and branching of this surface. As a result, it is presently preserved in the form of blanket-form relict deposits occurring as peaks of the hillocks.

Marine terraces are represented by two successively lowering steps towards the sea. The terrace at 6–8 m height above sea level is more consistent and more widely developed. Its formation took place during the late Pleistocene. The younger terrace is at a height of 1.5–2 m above sea level and was formed in recent times. A. Bateman considers its formation within an interval of 2100 to 2600 years.

Aden Rift

CENOZOIC STRUCTURE

As in the case of the Red Sea rift, the Aden rift, bordering the Arabian peninsula from the south, is characterised by a similar strike of the coastal line and its structural components. Its tectonic activity is analogous to the Red Sea rift and is characterised by mid-ridges, zones of transformed faults and an axial trough and is manifested in their high seismicity, elevated heat flow and accumulation of hot metalliferous deep-seated sediments. It differs from the Red Sea rift by a well-exposed mid-ridge. The Aden rift originates from the Tajur Bay, is 250 km wide in the west, increasing to 500 km in the east. The underwater mid-oceanic Sheba ridge, having a width of 100–150 km and intensively split-up axial furrows, is situated in the central part of Aden Bay. The Aden Median ridge is 240–280 km wide in the eastern part of the Bay at the Sokotra Island latitude and rises above the adjacent basin by 1900 m, rarely reaching up to 2500 m. The ridge is represented by a system of elongated ranges separated by deep valleys. The ranges reach up to 1500 m in height and 3–8 km in width. The angle of slope of the ranges exceeds 15°. A deep (~ 3000 m) trough-like valley with a flat bottom and steep slopes occurs in the centre of the Aden Median ridge. In distinction to the Red Sea rift, other characteristic features in the formation of the Aden rift are its highly reduced shelf and continental slope with a thin continental crust occurring in the form of narrow linear zones. Though the newly formed oceanic type crust in the Red Sea rift is developed only in the axial trough, the same is much wider in the Aden rift. Based on geophysical data, A. Laughton, J. Cochran and others consider its extension up to the continental slope.

(115)

Extensive development of transformed faults, having a north-west strike and 35° inclination, is characteristic of the composition of Aden Bay. About five-six major faults have been established on the basis of geophysical and bathymetric data. Besides these faults, I. Kalyaev and others have identified more than ten linear dislocations in the form of narrow, 2–3 km wide depressions in the relief at 50–150 m depth on the basis of data from specialised geologic-geophysical expeditions of the 19th voyage of the scientific-investigative ship 'Academic Vernadskii'. The morphology of these faults indicates their formation under a shearing environment. The larger transformed faults are associated with geniculated projections of the coastline and have undoubtedly the form of faults. The latter, like the Red Sea rift, were formed by the intersection of older faults; but in the Red Sea, their strike is north-west and submeridional while here it is north-east or sublatitudinal.

The more representative Alula-Fartak fault is situated along the straight line combining the African horn and the major protrusion of the Arabian Peninsula (Cape Fartak). The fault includes a 2820 m deep V-form trough, a narrow elongated range rising above the trough to 1700 m height and a reduced trough. The width of the fault zone is about 35–38 km. The components of the Alula-Fartak fault are represented by a step-fault structure.

The Aden Median ridge is displaced along the Alula Fartak fault to the SSW by 120 km approximately. The ridge narrows farther west and is morphologically less distinct and pinches out at 40° E lat. A narrow echelon type furrow, traceable in Tajura Bay and Lake Assal of the Afar basin, serves as its structural continuity. Change in the structure of the Aden Median ridge has been observed during its transition into major transformed faults.

The marginal faults of the Aden rift are older in origin and are associated with the formation of the Jurassic trough (Proto-Aden rift). However, the main stage of development of these structures has been in the interval from the early Miocene to the present. The internal faults of Aden Bay, associated with formation of the Aden Median ridge, were formed at a later stage of rift formation. They were tentatively formed in the late Miocene (around 10 million years) when the oceanic-type crust had already formed.

(116) In the opinion of A.F. Gracheva, the median ridges of oceanic rift zones are secondary with respect to the young oceanic crust [6]. They appear in the axial part of the rift zone on reaching a definite stage of shearing. The Red Sea and Aden Bay rifts illustrate such concepts well. Shearing in the Red Sea rift caused thinning of the continental crust. However, this shearing was moderate and was accompanied by considerable vertical movement. Apparently, rupture of the continental crust and new oceanic-type crust formation took place only in the axial part, i.e., in the narrow zone of maximum shearing (vertical displacement). Here the median ridge is in the embryonic stage of development. Its rudiments are represented by a chain of small-amplitude uplifts situated along the axial trough [9].

The typical median ridge formed in the Aden rift together with an increase in amplitude of shearing. Displacements along transformed faults began simultaneously with the increased shearing; as a result, the initial linear axial trough broke up into a series of en echelon segments, separated by transformed faults. Thus, the transformed faults are similarly related to the degree of shearing and, consequently, with the newly formed oceanic crust. For this reason, the transformed faults of the Red Sea should be considered only as the initial stage in the development of transformed faults.

Identification of the relationship of transverse faults of the Red Sea rift (A.V. Razvalyaev, E.N. Isaev, M. Shonfeld) and Aden Bay (V.N. Kozerenko, V.S. Larster, V.A. Selivanov) with their continental extensions is an achievement in recent years. It has been established that transverse faults of the Red Sea and Aden Bay pass into the continent, where the continuity of these faults is marked by prolonged (from Precambrian) development and deep-seated submergence. These data show the adaptability of rift formation to the structural anisotropy of the basement on which it developed and indicates the pre-existence of its structural set-up. It is now well known that this phenomenon is typical for zones of continental rift formation. Such information prompts us to consider that transformed faults inherit ancient (Precambrian) faults whereas, according to the concept of J. Wilson, typical transformed faults were formed only at the stage of formation of rifts from the oceanic crust. However, to assign a greater role to transformed faults in the structure of the African-Arabian rift belt is disputable.

The amplitude of horizontal displacement of transformed faults and the scale of total extension of the Aden rift are complex and inadequately studied problems. Their resolution lies in the more general problem of rift and ocean formation. With respect to the Aden rift, this means delineation of the mechanism of formation of the oceanic crust beyond the oceanic ridges. G.B. Udintsev, A.L. Yanshin, A.E. Schlesinger and others attribute limited significance to the role of a 'spreading' model in the development of the oceanic crust beyond the mid-oceanic ridges. In the opinion of A.V. Razvalyaev, total isolation of Aden Bay from the concept of a 'spreading' model is unrealistic, inasmuch as the Aisha continental block is situated along the strike of Aden Bay with an oceanic-type crust. This situation is avoidable according to the new model of formation of the Aden Bay crust outside its axial range proposed by Cochran [38].

Although the main shearing in the Aden rift took place in the median ridge, the role of transformed faults was great. Vertical movements played a significant role here together with thrust (transforming) movements and tensional shear. Besides major linear transformed faults, traceable in the continent, the Aden Median ridge branches into a system of intersecting uplifts, narrow valleys and troughs, forming thereby a complex structural configuration reminiscent of the 'chessboard' pattern typical of the Kenyan rift and characteristic for an environment of differential vertical movements.

MECHANISM AND KINEMATICS OF FORMATION OF THE RED SEA BASIN AND ADEN BAY

Although the majority of researchers studying the problem of time of formation of the Red Sea basin agree on the late Oligocene(?)-Quaternary periods, they nonetheless differ significantly with regard to the mechanism of this process and magnitude of shearing of the basin. The suggested mechanisms of formation of the Red Sea basin can be combined into two groups: 1) associated with the development of an arch (dome) following the model of H. Cloos and 2) due to moving away of the Arabian block from the African and the Somali blocks and anticlockwise rotation of the Arabian block by 7° to 9°.

Similarity in the configuration of coastline and boundary escarpments is characteristic for the Red Sea basin and Aden Bay. An analysis of the structural set-up of the Red Sea rift zone showed that this peculiarity is only an isolated case in the general symmetry of the rift zone and appeared because of the similarity of composition of its constituent components (axial trough, shelf steps, 'shoulders') with belts of alkali magmatism. Similarity in strike of coastline allows us tó presume that the coastlines were combined earlier, but the amplitude of expansion or separation, calculated from the suggested model of A. Laughton, D. Davies, K. Tramontini and others, works out to be 200–300 km.

The total expansion of the Red Sea, calculated from combining the coastal line up to 250 to 300 km, is not acceptable insofar as it reflects an extreme and unfounded mobilistic tendency and is not supported by many researchers, including supporters of the tectonic plate concept, such as D. Ross, J. Schlie, P. More, X.Le. Pission, J. Cochran and others. These concepts, as applied to the Red Sea and the Aden rift, were repeatedly subjected to strong criticism. E.E. Milanovskii in particular showed the (118) unsoundness of attempts to 'conjugate' the Aden Bay coast and the Aisha continental horst, which is the projection of the Somali block and which is located on the strike extension of the southern marginal trough of Aden Bay.

It has been suggested that the Red Sea and Aden Bay basins underlie an oceanic-type crust formed during two 'spreading' stages. As regards the Red Sea, many supporters of the 'pure' spreading model do not support this concept for its origin. According to geologic-geophysical data, the crust, which geophysically corresponds to the oceanic type, could be established only for the oceanic trough.

The amplitude of expansion (separation) of the Red Sea has been assessed as ranging from a few tens of kilometres to 250–300 km by different authors. Such a wide range undoubtedly reflects the absence of reliable criteria for its delineation and also a preconceived approach to specific theoretical concepts (in particular, the spreading model of formation of the Red Sea rift). As already mentioned, the concept that the newly formed oceanic-type crust occurs under the entire Red Sea basin does not conform to the geologic-geophysical data. The more reliable suggestion is that although the continental crust was subjected to shearing, its total separation from the newly formed oceanic-type crust was possible only in the axial trough and, consequently, the amplitude of expansion should not exceed its width, i.e., 30–50 km. The cumulative magnitude should take into account a scattered shearing along the entire basin both during the formation of the main trough in the Oligocene (?)-Miocene as well as in the Pliocene-Quaternary period, when tectonomagmatic activity in the rift was localised predominantly in its axial trough.

The concept of the presence of a newly formed oceanic-type crust under the entire Red Sea basin was further supported through the publication of I.R. Cochran [38] on the structure and evolution of the young oceanic basin and continental boundary, taking the Aden Bay as an example. Based on a study of the magnetic field in Aden Bay, the author established from the results that zones of neutral (quiet) magnetic field or loss of magnetic correlation exist between the Sheba Median ridge with the oldest magnetic anomalies corresponding to an anomaly of 5 (10 million years) and the continental shelf. These areas are up to 75 km wide and characterised by the weakly dissected relief of the bottom and lower gravitational field. The seismo-acoustic basement occurs at approximately the same depth within them and no inclination from the axis of the ridge could be detected. They are bounded by faults on the continental slope side. Magnetic anomalies, not older than 4–5 million years, have been detected in the western part of the bay. Consequently, in the opinion of I.R. Cochran, when the Aden Median ridge formed in the eastern part of Aden Bay during 10 and 5 million year intervals, the western part of the bay should have experienced shearing that differed from the spreading model in its mechanism. On this basis he concluded that the crust oc-

(119) curring beyond the axial ridge in Aden Bay was subjected to scattered shearing (diffused) and hence differs in origin from the crust of the Aden Median ridge.

Insofar as the formation of structures in the Red Sea and Aden Bay are interrelated in the set-up of the consolidated Arabian block, expansion in the Red Sea basin during the period of its formation (up to 10 million years) should have been scattered and should have taken place because of dyke intrusions and formation of normal and transverse faults. For example, as it is extensively manifested in the margin of the Afarian continental crust, the New Scotland Shelf in the southern Australian boundary zone with a quiet magnetic field and Bisco Bay, etc. I.R. Cochran [38] totally excluded the spreading mode of formation of the main trough of the Red Sea basin. He considered that the main trough, as in Aden Bay, was formed in the Miocene because of the same scattered expansion, which is operative at present in the northern part of the Red Sea. The author links shearing and formation of structure with the warming up of the upper mantle, i.e., with the development of endogenetic processes. Blocks of granitic rocks are known to occur in zones of a quiet magnetic field. The velocities of longitudinal seismic waves characteristic for a 'granitic' layer have also been established in Aden Bay.

The above-mentioned data help to reveal the nature and kinematics of formation of the Red Sea and Aden Bay basins. First of all, in the case of Aden Bay these data confirm the possibility of formation of the crust, having seismic velocities close to the velocity in the basalt layer, by a mechanism other than the spreading model. However, they bring no clarity to the scale of expansion. The initial standpoint of these concepts is formation of the basins due to horizontal movements. For example, the amplitude of expansion in the Red Sea at 17° N lat. constitutes 225 km, of which 80—85 km belongs to widening of the marine bottom related to the formation of the axial trough, and the remaining 140—145 km expansion took place because of scattered shearing during formation of the main trough of the Red Sea.

The above concept on the large-scale expansion of the Red Sea rift does not solve the problem as to how and where these movements took place along the strike of the western Arabian rift system, with which the Red Sea rift constitutes a dynamic pair. To date, there is no satisfactory explanation for the fact that if displacement took place from the south to north along the Levantin zone of faults (Western Arabian rift system), including the successively occurring grabens of Aquaba Bay, the Dead Sea, Tiveriad Lakes, Rhab and Karasy following the pattern of a left-sided thrust (Fig. 28), then, possibly one should not totally negate the concept that the dimension of amplitude of these displacements as presumed by researchers starting from A. Kennel (for example, 107 km for the Dead Sea graben) is not geologically sound. Let us mention the main arguments in favour of displacement along the Western Arabian rift sys-

(121) tem, which are shifting of facies and change in thicknesses of Cambrian, Jurassic and Cretaceous sediments and also the presence of zones of copper and manganese mineralisation along both sides of the Dead Sea, as established by D. Lubertre, A. Kennel and R. Freund. It may be fully explained from the position of the suture of the Red Sea and its location at the boundary between two blocks, differently directed in their mode of development, such as the Jordan block with a rising tendency in the east and the eastern Mediterranean pericratonic furrow in the west (Fig. 29). The thrust displacements along the Levantin fault (north Lebanon) with an amplitude of only a few kilometres remains unexplained from the viewpoint of major thrust movements.

The attempt to link shearing due to displacement with the formation of intraplatform fold zones (Palmira avlakogen and the northern Sinai zone of folds) in the frontal parts of moving blocks, namely the Arabian block in the east and Sinai-Levantin block in the west, is interesting. However, the structural configuration of deformations of these zones has not been recorded in the kinematic set-up, following which the structures in the limbs of displacements should have been mirror reflections. The folds occurring in the Palmirides as well as to the north of the Sinai Peninsula are characterised by similar structural types: the southern and south-western limbs of the anticline are, as a rule, steep and disturbed by faults, even by overthrusts. The general plan of virgation of the folds is also different in these zones. It must be noted that a situation of extensive shearing in the volcanogenic Druz massive in southern Syria and Jordan cannot be explained from the viewpoint of a left-sided slide of the Arabian

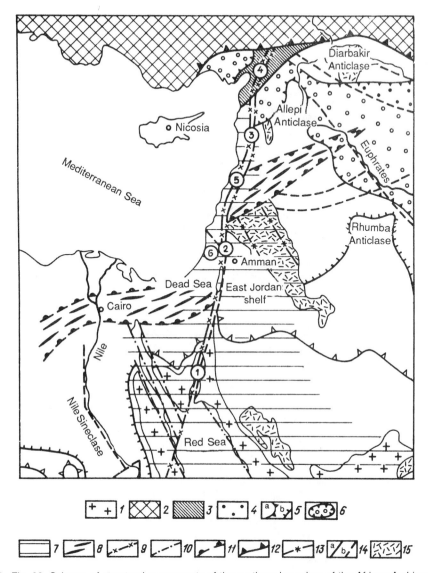

(120) Fig. 28. Scheme of structural components of the northern boundary of the African-Arabian rift belt (after V.P. Ponikarov supplemented by the Author).

1—Arabian-Nubian shield; 2—Alpine folded belt of Tavra region; 3—Marginal part of the Arabian platform reworked by movements due to the Alpine orogenic cycle; 4—Mesopotamian marginal trough; 5—anticlases and synclases, contour mapped along floor (a—upper Chalk, b—Eocene); 6—Neogene furrows; 7—Red Sea-West Arabian regions with epiplatform activisation; 8—axes of folds of 'Syrian arches'; 9—faults of western Arabian rift systems; 10 to 11—marginal faults: 10—Red Sea and Suez Bay basins, 11—'Syrian arches'; 12—overthrust zones of the Alpine folded belt; 13—feeder canals (faults) for lava flows; 14—major faults (a—established, b—assumed); 15—Neogene-Quaternery basalt sheet. Basins: 1—Aquaba; 2—Dead Sea and Jordan River valley; 3—Rhab; 4—Amuk and Karasu; upliftments: 5—Lebanon-6—Juda 'dome' of West-Arabian rift system.

122

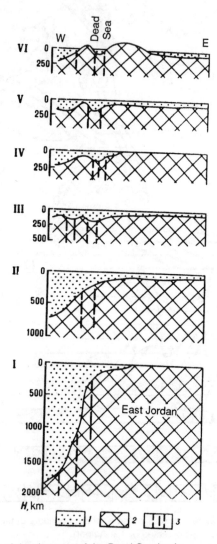

(121) Fig. 29. Scheme of development of the Dead Sea basin.

Palaeotectonic profile: I—early and middle Jurassic; II—early Cretaceous; III—late Cretaceous; IV—Palaeocene and early Eocene; V—middle and late Eocene, VI—Oligocene; 1—regions of sediment accumulation; 2—underlying substrata; 3—zones of synsedimentary faults.

(122) block along the Western Arabian rift system, insofar as the major basaltic eruptions in this region took place during the Oliocene-Quaternary period along the north-west-striking faults, which must have been subjected to

compression, i.e., 'grinding' following the kinematics of movements in the Arabian block.

Moreover, a few structural features in the Dead Sea region, observed by A. Kennel as well, undisputedly indicate the play of left-sided (shearing) tension along the Levantin rift system. These are overthrusts diagonally receding from the north-eastern and south-western ends of the Dead Sea and oriented in different directions. These anticlinal folds were recently established by I. Salateh in the Amman city region. They occur contiguous to the eastern fault of the Dead Sea. Finally, the left-sided shifts with an amplitude of up to 600 m occurred in the Quaternary period and have been recorded in aerial photographs.

A. Nur and Z. Ben Abraham showed that the actual tension of horizontal shearing in the Red Sea rift becomes dissipated and is dampened inside the Arabian block along a series of faults and intraformational (intraplate) fold structures of the Palmira inversional avlakogen type, which defines the Livantin rift system of faults, namely Azhlun (Jordan), Carmel (Juda Hills), Roshai, Sirhai and others (Lebanon). Consequently, the shearing tension is almost completely dampened inside the Arabian lithospheric block and does not reach the frontal part of the Alpine belt. Moreover, the role of horizontal compression from the side of the Arabian block and its active role in the formation of external chains of the Alpine belt (Zagros, Tavr) should not be denied. Underthrusting of the Alpine plate under the Alpine region was noted by V.P. Ponikarov even in 1967. However, in the light of the above statements, it has apparently been achieved through independent movement of the Arabian plate. The latter apparently moved 'along the shoulder' with the entire African continent. As regards the independent tension from shearing, which originated as a result of endogenetic development of the Red Sea rift zone, this got dispersed inside the Arabian plate. This fact once again places the reality of the existence of 'rigid' lithospheric plates in doubt.

Metallogeny of Red Sea Rift Zone

The practical significance of a study of the Red Sea rift zone arises from the theoretical basis of distinguishing the prerift stage and its specific magmatic formations, characterised by specific metallogenic specialisation. Thus, carbonatites with potentiality of rare metal occurrences are associated with deep-seated alkali-ultrabasic and basic magmas, magmatogenic copper-nickel and titano-magnetite ores are associated with alkali-gabbroid and gabbroid rocks, and rare metal and rare earth deposits and also deposits of gold, wolfram, molybdenum, tin, etc. are related to derivatives of crustal alkali-granitoid magmas. Detection of the patterns of spatial and chronological emplacement of these formations during the prerift stage has great significance for establishing the regional and local

(123)

control of the related manifestations of ore deposits, particularly in regions showing ancient rift formations [23].

Delineation of the Arabian-Nubian alkaline provinces, determination of their parameters, formational characteristics and metallogenic special-isation have great theoretical and practical significance. Carbonatite ex-posures associated with ring intrusives of alkali-gabbroid composition sig-nificantly expands the forecasting limits and exploration for ore-bearing carbonatites in continental rift formation zones, associated not only with alkali-ultrabasic, but also with alkali-gabbroid formations. The established patterns of the structural-kinematic localisation of ferro-manganese ore deposits and contemporary thermal and metal-bearing solutions in deep-water basins of the Red Sea have specific significance in the methodol-ogy for studying other continental rift zones and particularly their ancient analogues, i.e., palaeorifts (avlakogen), and contemporary oceanic ore genesis.

Ferro-manganese and lead-zinc mineralisation are associated with specific stages in the formation of the Red Sea basin. In the Red Sea rift zone manganese ore deposits are developed predominantly in the west-ern coast of the Red Sea. In this region they are localised in the Ras-Banas, Halaib and Tokar regions as well as in the northern part of the Afar depression (Fig. 30). Manganese deposits are situated in the north-ern part of the Arabian coastal regions of the Red Sea. Their occurrence in the western coast has long been known and they have been worked from the beginning of this century. However, the question regarding their genesis remains a debatable point. M. Kabesh, F. Atia, V.V. Balkhanov, A.V. Razvalyaev and V.V. Ishutin consider them to be of the hydrother-mal type whereas A. Bateman and others regard them as sedimentary. Obviously, the solution of the problem of origin of manganese ores is not only of theoretical but also of practical significance, insofar as the ap-proach to exploration methods depends on the acceptance of one or the other viewpoint. The theoretical importance of this problem increases if the hydrothermal origin of ores and their relationship with the formation of rift depressions are recognised.

All the manganese-bearing regions of the western coast of the Red Sea are characterised by their more or less similar structural set-up. The Khalaib region may be discussed as an example. Almost all the known ore bodies of this region (Fig. 31) are associated with fault dislocations. The fault systems to which they are confined have a western or north-western orientation (strike direction 315° to 330°) and are subparallel to the coast line of this part of the Red Sea. The fault dislocations are distinctly traceable in aerophotographs through disposition of numerous open-cast pits (trenches) oriented in a linear fashion or sometimes an en echelon pattern. Extension of these faults is measured in kilometres. The

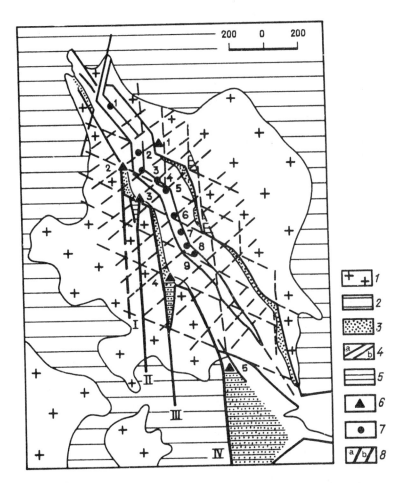

(124) Fig. 30. Scheme of structural control of iron-manganese deposits and deep water metalliferous basins of the Red Sea rift.

1—Precambrian rocks; *2*—Phanerozoic platform cover; *3*—Neogene-Quaternary deposits in rift basins; *4*—main rift faults *(a)* and network of diagonal faults *(b)*; *5*—triangular rift basins, Afara type; *6*—regions of iron-manganese deposits: 1—Arabian, 2—Ras-Banas, 3—Halaib, 4—Tokar, 5—Afara; *7*—deep-water basins of the Red Sea: 1—Oceanograph, 2—Kebrit, Gipseous, 3—Waldivia; 4—Nereus, 5—Tetus, 6—Atlantis, li Discovery, 7—Shagara, Erba, 8—Port Sudan, 9—Suakin; *8*—Submeridional rift faults *(a*—established, *b*—assumed): I—Ras-Banas; II—Diib; III—Barak; IV—Danakil.

fault dislocations are localised either in a wide coastal band (15 to 30 km), comprising coastal marine Neogene-Quaternary rock formations, or in (124) granites and andesites of late Proterozoic age. In both cases, the ferro-manganese ore deposits are represented by steeply dipping ore bodies occupying space between fault displacements of a north-west strike. Ore veins, localised in granites, represent monolithic massive bodies. Mineral aggregates have a coarse-grained texture. The ore zones attain 10–15 m in thickness whereas individual ore bodies are 1.5–2.5 m thick.

(126) The ore bodies in the coastal plain intersect the near horizontal Olio-cene-Quaternary beds at right angles. The Oligocene-Quaternary beds are composed of sandstones, siltstones with lenses and intercalations of fine to coarse pebble-bearing conglomerates and gritstones. In contrast to the ore bodies, localised in granites and having sharp contacts, these ore bodies show swellings (up to 2–3 m) and lensoid layers which bifur-cate from the main ore bodies and penetrate into the wall rocks at some places. They are represented in a few deposits by a series of contiguous iron-manganese ore veins, combined by radiating joints. The thickness of the ore bodies in the Pliocene-Quaternary sediments rarely exceeds 1.5–2 m.

The manganese content in these ores varies from 30 to 60%, with an average value of 40% and the iron content from 10 to 30%. The main manganese-bearing minerals are pyrolusite, manganite, psilomelane and rhodochrosite. These ores also contain barite, galena, hematite, pyrite, calcite, chalcedony, magnetite and goethite.

The morphological characteristics of these ore bodies are their oc-currence in the form of veins, confinement to the Precambrian gran-ites and intersecting (vertical) relationship with the associated subhori-zontal Pliocene-Quaternary sediments. These characteristics favour the hydrothermal origin of this iron-manganese mineralisation. It owes its ori-gin to emerging hydrothermal solutions, controlled by north-west-striking faults. The ore bodies were formed from low-temperature solutions un-der near-surface conditions and filled in the open spaces produced by fault displacements. It is interesting to note in connection with the char-acter of ore deposition that the hydrothermal solutions moved freely up to the surface, flowed around and occurred as 'envelope' around the late Quaternary granite and basalt pebbles.

The iron-manganese ore-bearing regions occupy a definite position in the general structure of the Red Sea rift zone (see Fig. 30). Their con-finement to geniculated bends of rift basins, caused by the intersection of major faults with submeridional and north-west strike, forming thereby wedge-shaped projections of rift structures into the continent is appar-ent. These regions are reminiscent of the embryonic triangle of the Afar Type and are characterised by variations in crustal structure, its thinning and possibly splitting. The iron-manganese deposits have an analogous structural set-up in the Red Sea coastal regions of Arabia and are mirror reflections of the manganese regions in the western coast. The similar-ity in the constitution and occurrence in space of the regions of iron-

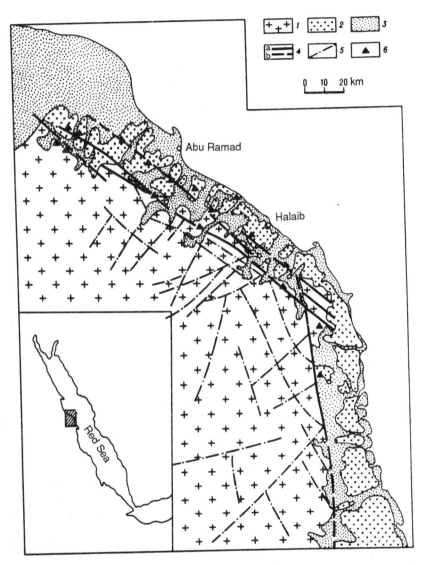

(125) Fig. 31. Geological-structural scheme of the Halaib manganese-bearing region (after V.V. Balkhanov and A.V. Razvalyaev).

1—Precambrian crystalline basement in the marginal uplift of the Red Sea rift (granites, andesites, basalts); 2—Neogene-middle Quaternary continental and lagoonal-marine sediments of the Red Sea rift basin, 3—Upper Pleistocene-Holocene continental and marine sediments of coastal plains; 4—main marginal faults of the Red Sea basin (a—clearly revealed in geological structure and relief, b—concealed under contemporary sediments); 5—other fault dislocations in the crystalline basement; 6—manganese ore deposits and ore occurrences. Location of Halaib region shown in inset.

manganese mineralisation may be considered as an indication of their genetic relationship with a single geodynamic process. The most important kinematic situations of the latter are characterised by vertical fault displacement of components which are inherent in the Red Sea rift zone. The vertical fault displacement of components associated with the general expansion of the Red Sea across its strike favoured exposure of the (127) triangular structure (depression) and penetration of the rifting process into the continent along the most predominant major submeridional fault dislocations, such as Ras-Banas, Diib, Barak and Afar. The shift component in the NNE direction favoured preopening of NW-striking faults. Thus, a combination of vertical fault-displacement vectors created a favourable kinematic situation for ascent of hydrothermal ore-bearing solutions. The northern component of the shift along the submeridional faults with up to 20 km amplitude has recently been confirmed by M. Shonfeld. Two fault systems with 315° and 330° strike occur in the Halaib ore-bearing region as well as in the entire Red Sea rift zone. Among them, the faults with the first mentioned strike are maximum ore-bearing and turned out to have maximum opening, whereas faults of the second system underwent less fissuring.

The Red Sea rift attracts the attention of a large number of geologists seeking to ascertain the nature of the contemporary oceanic ore formation process. Although the iron-manganese bearing nodules are the only objects of practical interest at present, the significance of ore genesis under oceanic conditions, particularly detection of submarine massive sulphide ores, is difficult to reassess. In this regard, study of the typical contemporary tectonic situations of oceanic ore formation and, in this context, first of all in the oceans under formation, i.e., in places where the structural-morphological relationship of oceans with continents is preserved to a great extent, assumes particular interest.

The Red Sea is one of the few classical examples of contemporary originating water bodies where one may trace the destruction processes of the continental crust from the initial to the final stages of the formation of an oceanic-type crust. As compared to already formed oceans, a study of the tectonic aspects of ore genesis in still forming oceans is particularly realistic for the methodologic plan of investigation, insofar as the geologic link of oceans with continents is not yet lost and may therefore be clearly traced and, consequently, may be presented as proof for the tectonic position of oceanic ore formation.

The uniqueness of the Red Sea lies in the fact that the 'classical' laboratory of contemporary ore formation in oceans is confined to its axial trough. They include deep-water basins with hot metal-bearing mud, such as Atlantis II, Valdiviya, Discovery and others. According to the data published by G. Pantota, such basins are 23 in number. Voluminous literature is available on metal-bearing solutions of the deep-water basins of the Red Sea, in particular the Atlantis II basin. The latter also served as the object of special investigations in the Oceanologic Institute. However, publications on metal-bearing sediments of deep-water basins deal mainly with the mineralogic and geochemical aspects of ore genesis.

The hot metalliferous solutions in the Atlantis II (60 km^2) basin are enriched in Fe, Mn, Zn, Cu and Pb. Four predominantly hydrothermal (lower—sulphide, central—oxide and upper—sulphide and amorphous silicate type) and detrital oxide-pyritic zones are segregated among them. The sulphide zones containing 5–20% zinc and 0.9–2.0% copper are of economic interest. H. Puhelt calculated in 1984 that the ore-bearing mud up to a depth of 25 m in the Atlantis II basin contains 30 million tonnes of iron, 2.2 million tonnes of zinc, 600,000 tonnes of copper and 6000 tonnes of silver. About 60,000 tonnes of zinc and other metals could be annually exploited over a period of 16 years.

(128)

It is necessary to emphasise from the tectonic viewpoint that deep-water basins of the axial trough are not irregularly distributed. They are, as if, 'fixed' in the framework of the fault network common to both the Red Sea water expanse and the bordering continents. Commensurable distance between deep-water basins of the axial trough and the north-west-striking faults in the continent helped us to see in them the manifestation of the fault network 'front' and commonality in their structural control. It is noteworthy that the structural bends of the axial troughs, in case they are not accompanied by known deep-water basins (lat. 25°00', 22°00' and 19°00' N), are also recorded in the fault network 'front'. They show indirectly the possibility of revealing deep-water basins and their accompanying hydrothermal processes in such places. It is also important to note that the distance ('front') between deep-water metal-bearing basins of the axial trough is approximately equal to 75–80 km, which conforms well with the intervals between transformed faults of mid-oceanic ridges.

The distribution of deep-water basins of the Red Sea is similar to the typomorphic situations of manifestation of submarine hydrothermal ore genesis established by P. Ron in 1984 for oceans. Ore genesis is localised in the latter as follows: 1) places of intersection of mid-oceanic ridges of the rift valleys with transverse faults, where perennial thermal sources were generated together with the formation of scattered disseminations, stockwork and massive sulphide ores and deposits of iron-manganese oxide-silicate ores; 2) similar areas but with thinner oceanic-type crust (stratiform and stockwork sulphide ores of copper and iron); and 3) places of bends in axial parts of the rift zones (massive sulphides). The tectonic situations for all the three types of localisation of oceanic ore genesis have been identified in the axial trough of the Red Sea, but only in 'embryonic' form. Moreover, there are deviations as well, namely, deep-water metalliferous basins are absent in the southern part of the Red Sea rift, where the axial trough is also intersected by transverse faults. Here the Red Sea rift has maximum conformity with the Precambrian substrata and the presence of ancient gneissic granite complexes may be presumed in it. Maximum development of metal-bearing basins is characteristic for the central segment of the rift, where the latter is sharply discordant with respect to the late Proterozoic Red Sea fold belt consisting mainly of volcanogenic rock complexes of basic and intermediate composition. Metals (129) might have been extracted here from the Proterozoic volcanic rocks including the island arc complexes, with sulphide ores, similar to the Umm-Samiuki ore deposit (Egypt). This may be interpreted in favour of the effect of the composition of substrata on ore genesis in oceans. Consequently, the structural and compositional heterogeneity of the Red Sea rift basement, although reworked (digested) in the axial trough up to the stage of formation of the oceanic-type crust, affects localisation of ore formation.

Thus, a commonality in structural control of iron-manganese deposits and deep-water basins of the Red Sea with hot metalliferous solutions of Atlantis II-type basins has been revealed. All types of mineralisation could be fixed in a fault framework according to the time of formation of the preceding rift formation. This helps us to see the uniformity in the genetic relations of ore genesis. The similarity in mineral associations of metalliferous muds of the deep-water Atlantis II basin and iron manganese deposits, particularly of the Halaib region, does not contradict the foregoing. The analogy becomes more convincing if we consider that the Halaib iron-manganese region and the Atlantis II basin are situated on the same fault dislocation. Thus, it may be considered that metalliferous solutions of the Atlantis II-type deep-water basins of the Red Sea and iron-manganese mineralisation of the Red Sea coast have a single deep-seated source, manifested in adverse structural facies environments. Ore genesis in deep-water basins is revealed in the form of metalliferous muds and in continents is manifested in the formation of iron manganese and lead zinc deposits.

Recognition of the hydrothermal nature of iron-manganese mineralisation in the Red Sea rift zone has great practical significance, insofar as it may guide geological prospecting for locating ore-bearing faults. The revealed patterns are also important for other rift systems, particularly for their palaeostructural analogues.

The following main structural features of the Red Sea-Aden rift zone during the Cenozoic stage of its geologic development should be stressed:

1. The leading role in the structural set-up of the Red Sea rift zone belongs to the orthogonal and diagonal fault network, which brings out the symmetry of formation and morphology of its constituents. These are

the axial trough, shelf zones, main trough and the bordering uplifts or 'shoulder' of the rift.

2. The faults (dislocations) are of ancient late Precambrian origin. They were subjected to repeated activisation during the Phanerozoic, giving rise to block formation of the Precambrian basement and controlled the emplacement of ring intrusives.

(130)

3. Submeridional and north-west-striking faults dominate the orthogonal and diagonal system of dislocations. They are observed to control the general morphology of the Red Sea rift zone. The intersection zones of these faults give rise to the geniculated bends in the Red Sea rift, maintaining its general north-west strike.

4. Faults with a north-east strike also had a significant role in the composition and formation of the rift basins. Their controlling role has been noted in the localisation of deep-water basins with hot metalliferous solutions. The latter are confined to the zones of intersection of these faults with north-western and submeridional faults, i.e., localisation of the basins is predetermined by the fault framework which preceded the formation of the Red Sea rift.

5. Junctions of the regmatic fault network are characterised by maximum contemporary tectonic activity. Deep-water basins with hot metalliferous solutions in the axial trough and zones of elevated heat flow and higher seismicity are confined to them. A similar assessment of the structural role of faults helped to explain the nature of localisation of the devastating Damar earthquake of December 13, 1982 (Yemen People's Democratic Republic). The delineated tectonic characteristics of the faults are significant in regional forecasting of earthquakes in the Red Sea rift zone.

6. The Red Sea rift was formed in the Oligocene(?)- Miocene; it was preceded by upliftment with insignificant gradient of relief during the Cretaceous-Paleocene. Formation of the rift was accompanied by short-periodic uplifts which did not exceed the Cretaceous upliftment in amplitude. Consequently, formation of the Red Sea rift does not fully represent the result of arch-formation, as conceived in the model of H. Cloos.

7. Structural evolution of the Red Sea rift clearly conforms to the two-stage development of rifts established by N.A. Florence and N.A. Logachev for the Baikal and other rift zones.

4

Correlation between Structural Features during Rift Formation and Prerift Stages of the Red Sea Rift Zone

(131) Chapters 1 and 3 have graphically presented the structural set-up of the Red Sea rift territory during the prerift and rift formation stages. Firstly, a discordant relationship has been distinctly established between the Precambrian and later Phanerozoic structural set-up of the territory with the purely riftogenic Cenozoic structures. The Red Sea rift has been laid on the pre-existing structural set-up; however, the degree of structural disparity differs for its individual segments, namely, northern, central and southern. Disparity of structural set-up of rift zones with their substrata ('frame') is a fairly typical phenomenon for continental rifts. The Red Sea rift is not an exception in this regard, but only confirms this aspect of rift formation. Besides, its study has helped to establish new peculiarities of the interrelationships of the structural set-up of the prerift and rift stages of development.

The correlation between the prerift and rift stages of the structural set-up in the Red Sea rift is more pronounced in relief in the central segment and may be illustrated using the Red Sea hills of Sudan as an example.

The structural inhomogeneities of the earth's crust, varying in dimensions, orientation and interrelationship between one another, have been revealed here on the basis of geophysical investigations. The zones of inhomogenous density and magnetic anomaly reveal a distinct similarity with the structural set-up established from geological data.

It was noted in Chapter 3 that the structure of the Red Sea hills of Sudan were defined as blocks based on geophysical fields (see Figs. 21, 22). The first order regional gravitational anomalies, such as the minimum for the Red Sea hills ($-82 \cdot 10^{-5}$ m/sec^2), the precoastal gravitational steps, the coastal maximum and the maximum for the axial trough of the Red Sea ($-140 \cdot 10^{-5}$ m/sec^2), clearly coincide with similar types of structural of features, namely upliftments ('shoulders') of the Red Sea hills marginal faults, pre-coastal steps, the main basin and the axial trough. These regional anomalies, commensurate with major structural components of the rift zone, were caused by rift-formation processes and reflect major inhomogeneities of the lithosphere which appeared in the Cenozoic stage of riftogenic activisation.

(132) The second order inhomogeneities in density, represented by regions of differentiated magnetic field and dividing the former into gravitational steps, have a north-east attitude and coincide with the late Proterozoic fold structures of the Sudanese-Arabian fold belt. Anomalies of this kind are non-coincident with the set-up of regional anomalies. The results of geological interpretation of the region show the relationship of these regions with changes in the composition of the earth's crust which took place in the Precambrian.

The regional gravitational anomalies of the Red Sea rift, appearing in the Cenozoic stage of rift formation, have a submeridional attitude. The interdependence of the positive maximum of the Red Sea and minimum of the Red Sea hills is revealed in the gravimetric field by the zone of intensive gradient of the Bouguer anomaly, which is related to an uplift of the Mohorovičić surface by 14 km. This zone coincides in relief with the eastern slope of the Red Sea hills, comprising Precambrian rocks.

The characteristic feature of the gravitational steps in the transition zone between the uplifts of the Red Sea hills and the precoastal plains is the fragmented nature of its structure. The gravitational step comprises a series of en echelon segments. The echelon type bends in the gravitational steps extend to places of its intersection with the gravitational steps of the second order which border the Precambrian blocks of the Sudanese-Arabian fold belt having a north-east attitude. Positive gravitational anomalies are confined to the zones of intersection of gravitational steps with north-east-striking gravity faults. These anomalies appear to have merged in the form of a wedge into the Precambrian structures of the Red Sea hills. Individual sides of these wedges have a north-east orientation which, apparently, attests to the regeneration of the Precambrian fault dislocations.

If it is considered that the gravitational anomalies reflect variations in thickness of the earth's crust as a result of deep-seated rearrangement of the lithosphere, then the positive gravitational 'wedges' should be considered as areas with thinned continental crust. These wedge-like areas, resembling the 'African triangle' in miniature form, are developed to a great extent in regions showing interdependence of major meridional faults with gravitational steps (see Figs. 21, 30), such as the Diib, Barak and Danakil faults.

A similar correlation may be interpreted as an adjustment of the riftogenic structures to the pre-existing Precambrian structural set-up. The endogenetic rift-forming process, causing deep-seated rearrangement, was superimposed on the Precambrian structural set-up and partially readjusted itself to the latter as a result of structural inhomogeneities and weakened zones of faults. The second order gravitational steps having a north-east orientation and also submeridional faults of prolonged development such as the Diib, Barak and Danakil, served as such zones.

The deep-seated impact of such an interrelationship is evident in the reworking of the lithosphere and its thinning. The rift-forming endogenetic process, as it were, 'ate away' the consolidated crust along the weakened zones (fault dislocations). The intensity of the process increased (133) from north to south. The coastal gravitational maximum, which complicated the minimum values for the Red Sea hills, is distinctly revealed in the submeridional segment of the Red Sea rift.

Geologic-geophysical data furnished on the Red Sea hills of Sudan and Red Sea may be interpreted as a reflection on the development of the Red Sea rift zone of two structural patterns, such as prerift and rift, that are characterised by their inherent features, which are manifested not only on the surface but also at depth. These data also reveal the non-coincidence of prerift and rift stage structural patterns and the superimposed character of the latter. The deep-seated rift-forming process transformed the consolidated crust after acting upon it and at the same time also adapted to the structural anisotropy. All these factors help in considering the rift-forming process as fairly independent and in observing the presence of sufficient independence in its formation. In the light of the Baikal rift, a similar interrelationship of rift formation with the preceding structural plan has been convincingly demonstrated by S.I. Sherman, S.M. Zamaraev and others. This pattern has been delineated by E.E. Milanovskii in the case of faults of continental rifts as a whole.

The Red Sea rift is localised in the late Proterozoic fold belt. The majority of researchers acknowledge the affiliation of most continental rifts to the structures of ancient Precambrian formations. This viewpoint was first expressed by F. Dixie and R. McConnell as applicable to the African-Arabian rift belt and was subsequently developed in the publications of the Soviet East-African Expedition, i.e., by N.A. Logachev, E.E. Milanovskii, N.A. Bozhko, E.A. Dolginov and others. According to this concept, continental rifts succeed ancient mobile belts formed in the Archean and termed tafrogenic lineaments by R. McConnell [44]. A characteristic feature of the lineaments is their repeated tectonomagmatic activation. The tafrogenic lineaments and their accompanying rift faults actually reveal coincidence in plan. Both of them 'avoid' the ancient structure of shields, for example, the Tanganyikan rift bordering the latter from west to east. An analogous relationship undoubtedly reflects the dependence of rift formation upon major structural inhomogeneities of the lithosphere and its inheritance from them. It also shows that the old cratons are, on the whole, 'unfavourable' for rift formation. Besides, in the opinion of N.A. Logachev, these concepts practically exclude the possibility of regeneration and superimposition (autonomy) of rift formation and do not bring out all the diversities of its manifestation. A more constructive approach in the study of the interrelationship of rift structures with their foundation is possible on the basis of the concept of the deep-seated energy potential of rift-forming processes [19]. According to this concept, the formation of a structure during rift genesis, its dimensions, intensity, magmatism and other characteristics depend upon deep-seated endogenetic regimes, i.e., upon the degree of heating up of the upper mantle in the prerift stage.

(134)   A similar approach to the study of the Red Sea rift zone helped the author to propagate a prerift or preparatory stage (according to the concept of N.A. Florensov and N.A. Logachev) and an earlier stage of the thermal stimulation of the mantle, manifested a few tens and even hundreds of million years before the main paroxysm of rift formation and conforming to the earliest 'embryonal' (according to E.E. Milanovskii) endogenetic regimes in the rift-forming regions. From these viewpoints, as shown in Chapters 1 and 2, the Red Sea rift zone is confined to the largest lithospheric inhomogeneity in the zone adjoining stable and mobile parts of the Central African craton. Apparently, due to the specific physical properties of the link zone between the two major lithospheric blocks, which developed during the late Proterozoic during different tectonic regimes, they became areas of manifestation of thermal stimulation of the mantle in the Palaeozoic and Mesozoic, as if to prepare the lithosphere for the subsequent rift-formation process. Consequently, the location of the Red Sea rift is, on the whole, predetermined in the regional plan by a deep-seated process. Concrete structural morphologic revealation of the rift depends, to some extent, upon the anisotropy of the Precambrian basement.

The Red Sea rift zone, inherited from the Precambrian lithospheric inhomogeneities and preceding the endogenetic regime (Palaeozoic, Mesozoic), reveals indications of superimposition (regeneration). At this stage, the degree of its manifestation changes sharply along the strike. Maximum superimposition (discordant relationship) of the Red Sea rift has been observed in the central segment, where it intersects the Precambrian structures of the Sudanese-Arabian fold belt at a large angle. The Cenozoic rift formation is more intensive in this part of the Red Sea rift; maximum concentration of deep-water basins with thermal and metalliferous solutions, epicentres of earthquakes, network of transverse faults and maximum width of the axial trough confirm this postulate for this region. At first sight, such correlation appears to be paradoxical and not to fit into the framework of prevailing concepts. However, it becomes clear if one considers that the Red Sea rift has inherited the prerift endogenetic regime, namely in the central part.

A similar interrelationship may be considered as an indication of the interrelationship of rift formation and its substratum at varying levels—mantle-based and crustal, among which variations are possible. In the case of the Red Sea rift, the north-east strike of the structural elements is the determining factor and formed during the entire history of the prerift stage of stimulation of the mantle, i.e., a deep-seated mantle-based process formed the sharply superimposed and, to some extent, independent rift structure of the Red Sea and, so to speak, 'removed' the less deep (crustal) anisotropy of the 'framework' (substratum).

The problem of distant terminals of rifts occupies a specific consideration in rift formation, insofar as their structural and morphologic interrelationships with the foundation are distinctly revealed in regions showing (135) rift thinning. It is known that the Red Sea rift ends in the north-west in the Suez graben and Gulf of Aquaba graben. The first one is the structural continuity of the Red Sea rift and is traced farther north-west up to the pericratonic trough of the eastern Mediterranean sea. However, only the coincidence of the strikes is common between the Suez Bay graben and Red Sea rift. Wide shelf zones of the main trough and the bathymetric elements of the Red Sea bottom appear to be cut off by the south-western continuity of faults of the Aquaba Gulf-Red Sea region or the Western Arabian rift system (see Figs. 19, 28). Faults in the Aquaba Gulf-Red Sea region are of ancient Precambrian origin and they demarcate two major lithospheric blocks with different tectonic regimes and deep-seated structures, such as the Jordanian block having a rising tendency to the east and the eastern Mediterranean (Western Arabian pericratonic trough) block—to the West (see Fig. 29).

Development of the Western Arabian rift system is associated with formation of the eastern Mediterranean depression. Of course, the Western Arabian rift system conforms to the major lithospheric boundary at depth and was capable of effectively influencing the growth of the Red Sea rift zone in a north-west attitude just as the submeridional Olekmin and Tyrkandian faults hindered growth of the Baikal rift zone towards the east, following the boundary of the Aden Shield as well as along the deep-seated Stanov fault. It would, however, be wrong to consider this as a hindrance to longitudinal expansion of the Red Sea rift only in this direction inasmuch as the rifting process itself was observed to be capable of totally 'restructuring' the Precambrian Sudanese-Arabian fold belt, being located approximately along the same azimuth interrelationship with the rift as in the case of the Aquaba Gulf-Dead Sea faults.

Apparently, the composition of the basement influenced the lateral growth of the Red Sea rift along with lithospheric inhomogeneity along

the Western Arabian rift system. The fact is that the role of late Proterozoic granites in the Precambrian basement rocks of the Sinai Peninsula
and the northern part of the eastern Egyptian desert increases sharply
whereas the more ancient granite gneisses predominate in the basement
rocks of the Aquaba Gulf coast. Most probably, the same characteristics are manifested in the north-western extremity of the Red Sea rift
that are observed in other rift zones, such as in the southern end of the
Kenyan rift [19] and north-eastern boundary of the Baikal rift zone. These
characteristics are: tapering out of rift structures in the course of their
emplacement in ancient cratons (Kenyan rift), or deviation from the initial strike along the periphery of these ancient cratons, as is detected
in the northeastern extremity of the Baikal rift, occurring in the proximity
of the Aldan shield. In the case of the Red Sea rift, the stable granite
saturated Sinai block appears to have fragmented the rift, forcing it to
accommodate itself to the structure of the basement and to utilise therein
the ancient structural inhomogeneity—the Aquaba Gulf-Dead Sea faults.
(136) However, neither these faults nor the stable Sinai block could totally stop
the massive growth of the Red Sea rift, such as the Baikal rift, and only
weakened it significantly. Therefore, the northern extremity of the Red
Sea rift is represented by the significantly less pronounced appendix—the
Suez graben. The Suez graben extends up to the Cairo-Suez fold zone
which is intersected almost at a right angle. Although there is a hypothesis in favour of northern continuation of the Red Sea and Suez Bay faults
within the Aegian Sea and Balkan Peninsula, this graben appears to be
incapable of further transforming this zone.

Based on the constitution of the Red Sea framework, the southern
extremity of the Red Sea rift seems to be related to the composition of the
Red Sea Precambrian basement. The Red Sea rift gradually tapers out in
a south-east direction and finally terminates in the narrow Bab-El Mandeb
strait, connecting it with Aden Bay. The Red Sea depression gradually becomes gentle in this direction and the axial trough becomes regenerated
instead of extending up to the Bab-El Mandeb strait. Tair, El Hanish El
Kabir, Zukar and other islands emerged here and in contrast to the typical
tholeiitic type volcanism associated in the axial trough of the mid-oceanic
ridges, here volcanism is transitional to the alkaline type, which is characteristic of the continental rifts. All these phenomena display weakening
in the growth of the Red Sea rift in this direction. As in the north-western
extremity of the Red Sea rift, its expansion in the south-eastern flank apparently hindered the advent of the rift in the granite-gneissic block. In
recent years the South-Arabian block in the east has been proved to be
composed of ancient granite gneiss. Distinction of such granite-gneissic
blocks is difficult in the west because of extensive development of basaltic
fields in the Afar depression and Ethiopian plateau. Exposures of granite

gneiss in these regions are, however, detected along the flanks of the rifts in Eritrea. V.G. Kaz'min assumed the presence of an ancient gneissic basement under a considerable part of the Ethiopian plateau basalts. Structural transformation of the Kenyan and Ethiopian rifts and their contiguity took place in the region lying to the north-east of Lake Rudolf where ancient granite-gneissic formations are exposed. This was first noted by N.A. Logachev and later by E.A. Dolginov.

About the south-eastern extremity of the Red Sea rift, it is conceivable that although the axial trough of the rift tapers out in the south-eastern direction, the amplitude of average stretching is compensated by faults with vertical movement in the Afar depression, particularly in the case of volcanic ridges confined to axial regions. According to T. Kristiansen, overall amptitude of stretching gave rise to emplacement of basalt dykes, shear fracture, and small scale faults in the central part of the Afar, amounting to 15 km. Results of geologic-geophysical investigations show that the continental crust in Afar is thinned out to 16 km though not torn apart. Crustal rupture is conceived only in the narrow axial volcanic ridges of the Erta-Ale type in the northern part of Afar in the Danakil depression.

(137) If it is considered that the Red Sea rift separates the Tazura and Danakil continental blocks from the Afar depression, then isolation of the Red Sea rift and correctness in considering its south-east extremity in the light of the concept given above become apparent.

During the stages of formation of the axial trough, the Red Sea rift represented itself to be sufficiently isolated with respect to its growth in the central part with gradually decreasing intensity of growth in the flanks. According to E.E. Milanovskii, this feature of the Red Sea rift once again confirms discontinuity in the rift-formation process along the strike of rift belts, or discrete development of linear platform structures and development of rifts from their centres to the flanks [21].

Continental rift genesis had notably developed in the Precambrian metamorphic fold belts and avoided ancient cratons. In 1973, from the example of the coastal Atlantic in South America and Africa V. Fyfe and O. Leonardos showed that rift formation followed ancient mobile belts with a complex metamorphic history. These belts characterise in granulite facies the metamorphism of rocks and their association with charnockites and migmatites. Their metamorphism is of the Abakum type with 60°C/km or more thermal gradient. The Trans-Amazonian belt (2000 million years) is an example of this type in the South American coast.

In distinction to cratons, the mobile metamorphic belts represent more heated structures of the earth's crust. Difficultly fusible granulite-charnckite-migmatite complexes were formed in the process of their evolution. Variations in the constitution of the cratonic crust and mobile metamorphic belts account for their different reaction to thermal anomaly which arose because of ascent of basalt columns during mantle diapirism or convective nuclei. Development of such an anomaly under an ancient craton with a thick granitic crust can be subdivided into four stages. Cracking of the crust along the initial weak zones took place in the first stage with intrusion of basalt dykes and eruption of basalts on the surface. In the second or maturing stage, a low viscosity layer emerged in the lower part of the crust due to its melting in the process of progressive heating; migmatisation and generation of granitic magma sources took place in it. This layer obstructed basaltic eruptions subsequently. In the third or concluding stage, or the stage of flow (spread) of the mantle diapir and its differentiation into the upper basaltic part, an anorthosite-perdotitic series of rocks formed. Granite formation ceased in this stage. In the fourth (fading) stage, thickening of the cratonic crust took place at the cost of the basaltic material of the 'extinct' diapir. Consequently, the craton was permeable for basaltic melts only in the initial stages of the mantle diapirism process. Subsequently, the craton underwent shearing during the course of spreading of the diapir. It did not, however, split up because of the considerable thickness and viscosity of the lithosphere.

(138) The mantle diapir developed under the mobile belt following approximately the same pattern with the principal difference that the presence of difficultly fusible granulite-charnockite-migmatite rocks at the base of the crust did not favour formation of a heated low viscosity layer and thus basaltic eruptions were obstructed. Splitting of the core with the formation of rifts is more feasible in the course of spreading of the diapir.

The suggested model explains why the ascent of the mantle diapir or thermal convection into ancient shields led to the emergence of dykes and vast trap regions; but the process does not advance further and does not explain complete splitting and formation of rifts, although the early stage of development of the mantle diapir is a destructive form of tectogenesis in nature. Parana, Karru and other traps were possibly formed at this stage. With the appearance of the low viscosity layer subsequently in the lower parts of the sialic layer, a minimal destructive process was attained. This model apparently explains why the rifts 'avoid' old cratons and why ancient mobile belts become the arena for future 'rifts'. In other words, this model puts forward the concept of a deep-seated mechanism of 'selectivity' of rifts.

V.E. Khain, F. Fyfe, O. Leonardos, N.A. Bozhko, E.A. Dolginov, N.A. Logachev and others have observed that rift formation was confined to unique late Proterozoic belts which underwent repeated tectonomagmatic activisation in the type areas of Mozambique, Atlantic, Grenville, Aravalli, etc. N.A. Bozkho considered them unique belts of tectonomagmatic transformation. The characteristic features of these belts are their polycyclic development under a high temperature regime with predominance of basitic (granulite-basitic) profiles of supracrystalline complexes and persistent inheritance of the infrastructural strike of the basement. Fold structures being formed, these belts underwent repeated tectono-magmatic transformation accompanied by tectogenesis (repeated folding, thrust formation, formation of schists and zones of increased deformation, etc.), retrograde metamorphism, metasomatism, granite and pegmatite formation and 'rejuvenation' of rocks. A persistent tendency to periodic uplifts and erosion, favouring exposure of the granulite-basite substratum, is manifested in the history of their development. Precambrian belts of a similar type are characteristic for the riftogenic passive boundaries of the Indian and Atlantic Oceans. The specific feature of the Precambrian belts of tectonothermal transformation is represented by the tendency to destructive type of development which is preserved even in much later stages of geologic history. Specifically these belts became the arena of Mesozoic and Cenozoic rift formation with which formation of younger seas was associated.

The late Proterozoic Red Sea mobile belt, situated in the northern continuation of the Mozambique activised belt, differs from the continental marginal belts in constitutional and developmental characteristics. As (139) was seen in Chapter 1, it is intracratonic in nature and belongs to completion of concluding stages of structural development. the Pan-African or the Mozambique orogeny was not activised in the Red Sea mobile belt, insofar as the tectonomagmatic events synchronous with this orogeny were 'fitted' into the natural cycle of development of the mobile belt, in this case the Baikal belt. Moreover, the activisation processes were 'directed' from this belt at least into the adjacent regions. Considering that in dimensions and extent of development the Red Sea rift excels all known continental and intracontinental rifts, it may be assumed that predisposition of this territory to rift formation was optimum. All these factors show that development of rift genesis is not restricted only to structures of the continental marginal belt type with their inherent granulite-basite complexes and that rift formation does not in the least depend, firstly, upon the structural inhomogeneities at the lithospheric level (Red Sea, Baikal and other rift zones), and secondly, upon the character of activisation in the prerift stage. Both these situations are manifested in the Red Sea rift zone.

The model proposed above for development of the mantle diapir is applicable in explaining the tapering of the Red Sea rift. Volcanogenic sedimentary rocks of basalt and andesitic composition and metamorphosed under green-schist facies conditions occur in the central part of the rift as a component of its Precambrian substrata. Herein, rift formation is more developed in the central part, where the rift has maximum disharmony with the structural set-up of the volcanogenic-sedimentary substrata, as

well as where the latter are weakly manifested in the northern part because the strike of the rocks of this substratum is highly conformable to
the rift. This fact undisputedly shows that rift formation is controlled, first of
all, by deep-seated phenomena and not solely by the structural features
of the substrata. Moreover, the substrata of the Sudan-Arabian belt do not
play the final role insofar as they are apparently closer to schist-basite or
mafic-type crust in composition (andesites, basalts), and therefore more
predisposed to riftogenic activisation.

The following conclusion may be drawn on the basis of the aforesaid statements. The Red Sea rift developed as an endogenetic active
system with its own centre of growth. The correlation among its prerift
and rift-stage structural set-up is manifested at different levels, such as
the mantle and the crust. At the mantle level, this correlation is manifested in the inheritance of 'the body' of prerift activisation of Precambrian lithospheric inhomogeneities by the endogenetic rift regime. From
this viewpoint the Red Sea rift is a pre-existing structure. The correlation
of the endogenetic regime as an essential rift stage with the basement
structure reveals superimposition (autonomy) of rift formation, wherein its
degree of manifestation changes within wide limits and depends upon the
anisotropy of the rift basement (frame).

5

Evolution of the Earth's Crust in Active Continental Rift Forming Zones

(140) The prevailing concepts in theoretical geotectonics and the extremely contrary expressions on fixation and mobilisation have imparted a new impulse to the concepts on the dynamic processes in the evolution of the earth's crust and the lithosphere as a whole. Publications by Soviet tectonists, such as A.V. Peive and his school in the Geological Institute of the Academy of Sciences, USSR, V.V. Belousov, V.E. Khain, Yu. M. Pushcharovskii, E.E. Milanovskii and others have shown that constructive-destructive processes constitute the base of the evolution of the earth's crust. The ideas on constructive development of the earth's crust found their reflection in the new methodological approach to compilation of a tectonic map according to periods of formation of the mature continental crust. Following this method, A.V. Peive and others compiled a tectonic map of Eurasia, A.S. Perfil'ev of the Urals, M.S. Markov and others of northeast USSR, E.D. Sulidi-Kondrat'ev of Africa and Arabia. In the opinion of V.E. Khain, Yu. M. Pushcharovskii, E.E. Milanovskii, L.E. Levin, M.S. Markov and others, the destructive tectogenetic process lies at the base of the contemporary concepts on the formation of marginal and internal seas and newly formed oceans.

At present, more and more researchers are inclined to believe that continental rift formation and ocean formation are the extreme steps of the singular destructive process. From this viewpoint, the rift systems evoke special interest today insofar as they, firstly, participate as destructive forms of tectogenesis and, secondly, their study helps to observe the initial stages of rift formation (ultimately the ocean genesis) within the continents [32]. Therefore, study of continental rift-formation processes evokes much interest, particularly in those cases where these rifts are directly related to oceans. The African-Arabian rift belt and, first of all, regions adjoining the Red Sea, Aden, and Ethiopian rifts (the Afar depression) are, in this respect, a classic place of manifestation of the newest and contemporary destructive tectogenesis. Here, it is still possible to trace the processes of destruction of the continental crust from the initial to the final stages with the formation of an oceanic-type crust. Therefore, it is not accidental that the Kenyan and Tanganyikan rifts and the Afar depression have served as objects of special investigations within the framework of the International 'Upper Mantle' project.

The western (Tanganyikan branch) and particularly the Kenyan (Gregory rift) have been fairly well studied by V.V. Belousov, E.E. Milanovskii, N.A. Logachev and others. Therefore, let us examine the structure of the Afar rift which differs from typical continental rifts by its combined role (141) between the oceanic and continental rifts. Based on data from geologists abroad and drawing on the author's investigations of the Red Sea and Western Arabian rifts, the Afar structure is considered from the viewpoint of a destructive process. Herein, only the most generalised information is cited because more circumstantial evidence on different geological problems of the depression is contained in many publications by researchers abroad, primarily P. More, G. Taziev and F. Barbery who have made major contributions, and also in the two-volume compilation specially devoted to the problems of the geology of Afar edited by A. Pilgera. Information on volcanism and tectonics of the depression are given in Soviet literature published by N.V. Koronovskii, A.F. Grachev, V.G. Kaz'min, E.E. Milanovskii, N.A. Logachev and others.

Structural Features of Destruction of the Afar Continental Crust

The Afar depression is situated at the trijunction of the Red Sea, Aden and Ethiopian rifts (Fig. 32). In plan, it is triangular in shape and represents an important junction of the African-Arabian rift belt. Conjugation of the structural trends of the Red Sea (NW-SE), Ethiopian (NE-SW) and Aden (E-W) rifts takes place here.

The Afar, bounded on all sides by continental blocks, represents a tectonic depression filled with Miocene-Pleistocene marine and lacustrine sediments. Miocene evaporite strata are extensively developed in the northern part of the basin and consist of gypsum, anhydrite and intercalations of rock salts whereas the main part of the depression is filled with sheet basalts of Miocene to Recent age. The oldest basalt sheets exposed in Afar have been assigned 25 million years of age determined radiometrically. The major part of Afar is constituted of stratified series of the Afar and Aden volcanites of Pliocene-Pleistocene age.

The Afar series is represented by monotonous subalkaline basalt layers alternating in upper parts with rhyolites and ignimbrites. The accumulation of volcanites in this series took place during an interval of 8 to 1.5 million years. In the marginal parts of the depression, Miocene volcanites

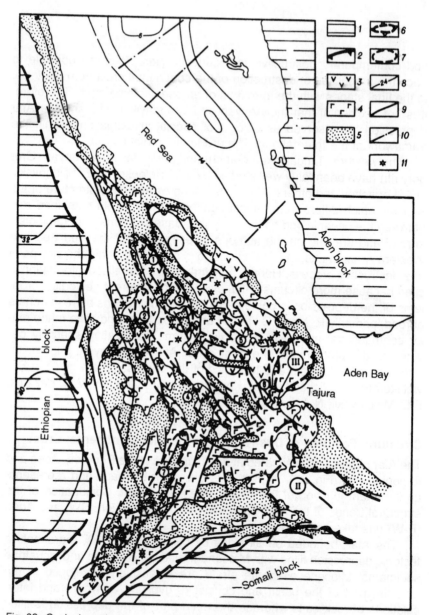

(142) Fig. 32. Geologic and structural set-up of the suture region of the Red Sea, Aden and Ethiopian rifts.

1—continental crustal blocks; 2—rift boundaries; 3 to 4—volcanic rocks: 3—Afar (Miocene-lower Pleistocene), 4—Aden (Pleistocene-Holocene); 5—red coloured and evaporite series of Neogene and marine, lacustrine and proluvial Quaternary deposits; 6—weakly 'transformed' blocks of continental crust: I—Danakil, II—Aisha, III—Tajura; 7—zones with maximum destruction of continental crust and ascent of thinned mantle: 1—Erta-Ale, 2—Alaito, 3—Amarti, 4—Tendaho, 5—Manda-Inakir, 6—Assal, 7—Issa, 8—Herta-Ale; 8—isolines of depth of occurrence of Mohorovičić surface, km; 9—faults; 10—transverse faults of the Red Sea rift, 11—volcanoes.

lie on older formations with an unconformity. The volcanites are interlay-
ered with lacustrine deposits. A swarm of dykes are distinctly traceable
within the sheet basalt of this series and were the conduits for lava flows.
The dykes in the southern part of the Afar depression show mainly NNE
and, to a lesser extent, NW and latitudinal strikes.

(143) In contrast to the basalt of the Afar series, the volcanites of the Aden
series are localised in narrow en echelon zones, giving rise to volcanic
ridges comprising basalts, trachytes, alkaline rhyolites and commendites,
which were erupted from central volcanoes. Petrochemically, they conform to the products of an oceanic type of volcanism. Their emergence in
the Afar series characterises a new series of expansion, which started in
the early Pleistocene and has been continuing to the present. According
to F. Barbery and J. Varrete, this phase of expansion started 3—4 million
years back and coincided with the opening phase of the axial troughs of
the Red Sea and Tazura Bay of the Aden rift. As a result of this process,
an oceanic-type crust was formed in the Red Sea and Aden Bay rifts.

The bottom of the Afar rift exhibits complex structure. A dense network of differently oriented faults with amplitude measuring a few tens and
rarely hundreds of metres (Issa graben, 500 m) is characteristic of it. A
considerable number of faults belong to types of open (gaping) fractures
with no displacement. Faults of larger dimensions give rise to differently
oriented systems of horsts and grabens, the largest among them being
the Ertale, Issa, Tendaho grabens, etc. Besides these forms of manifestation of tectonics in the Afar structure, there is a distinct system of axial
volcanic ridges which are confined to the zones of intensive stretching
with possible disruption of the granite-metamorphic layer.

The gravitational field of the Afar basin, on the whole, is characterised by a negative Bouguer anomaly. The axial volcanic ridges are
traced by zones of local gravitational maxima. The deep-seated structure
of the Afar depression and its adjoining areas, as revealed by gravimetric
and seismic data, shows that the continental crust is thin under the Afar
basin and is underlain by a layer of lesser density (7.3—7.7 km/sec velocity of longitudinal seismic waves). This layer, occurring at 16 to 24 km
depth in the Afar, has been interpreted from the magmatic-telluric data
of most researchers as the anomalously heated mantle, providing high
heat flow and generating fusion of weakly differentiated basaltoid magma
feeding the volcanoes of the axial zone. This zone has been identified
by F. Barbery and J. Varrete as the mid-oceanic ridges from the nature
of its volcanism and geodynamic regime. The axial volcanic ridges are
confined to areas where the thin mantle layers are nearest to the surface
(see Fig. 32). This is particularly distinctly revealed in the Erta-Ale volcanic chain in the northern part of Afar, where maximum thinning of the
continental crust or its total absence has been suggested.

Thus, we have a complex structure in Afar consisting of multidirectional faults, dyke swarms and axial volcanic ridges. All these reflect a complex interrelationship of tectonic stress fields leading to changes in the continental crust—its fragmentation, thinning, stretching and piercing by basic magma, i.e., destruction. The initial stages of this process are represented by intensive fracturing of the crust and intrusion of dykes, i.e., by its mechanical separation. Results from computations of the intensity of faults, fractures and dykes in the central Afar helped us to ascertain the (144) amplitude of stretching. The latter constituted 14.5 km for the period late Miocene-Recent (1500 faults, dykes and fractures were measured). According to M.G. Lomize, an analogous process of stretching of the crust with concomittant enrichment with basic dykes is characteristic of Iceland.

V.G. Kaz'min and others consider that besides direct faulting with vertical displacement and shifting of the crust, which are mostly exhibited in the axial volcanic ridges of Afar, destruction of the continental crust, i.e., thinning of its granite-metamorphic layer, may have taken place under intensive crust fracturing and one-sided inclination of the blocks [13]. This mechanical process in the stable consolidated crust at depth may combine with plastic deformation favouring the tensional process. O. Charpal and others have drawn a similar mechanism to explain thinning of the continental crust in general for the continental marginal rift of Biscay Bay and A.V. Peive for the Atlantic.

In addition to the observed processes, destruction of the continental crust took place due to saturation of the granitic-metamorphic layer by basic magmatic rocks, i.e., by dyke and sill-like injections which, in the long run, could have led to changes in the geophysical characteristics of this layer, bringing it, according to these parameters, to a 'basaltic' type. Numerous swarms of basalt dykes in Afar provide indubitable confirmation of this process; their density in some places is so great that the surrounding substratum has limited significance. This situation can be illustrated with north-eastern Sudan as an example, where the author observed a system of dolerite dykes in the Red Sea framework, sometimes completely replacing the Precambrian substratum. This apparently explains why the crust under the shelf stages in the main trough of the Red Sea has velocities corresponding to longitudinal seismic waves of 6.97 km/sec, which are inherent in a 'basalt' layer.

The possibility of saturation of the granite-metamorphic layer by basaltic material in the shelf zone of the Red Sea rift has been visually confirmed with the Sudanese Red Sea coast as an example. Here the author studied the sheet basalts among the mid-Miocene sedimentary rocks which filled the Red Sea depression at Cape Abu-Shagara 21° N lat. The rocks are petrochemically similar to basalts dragged from the axial trough, Red Sea islands and mid-oceanic rifts. The basalt eruptions took place about 25 million years back, i.e., much before the formation of the Red Sea axial trough (4–0 million years), and correspond to early stages of destruction of the continental crust. The appearance of practically undifferentiated basalts of the oceanic type with a formerly known continental substratum in the marginal part of the Red Sea depression is notable from the viewpoint of destructive tectogenesis. It reveals the high energy potential of an endogenetic regime providing a high rate of heat flow and intensive melting of the mantle, giving rise to a weakly differentiated tholeiitic basaltic magma.

(145)   It becomes apparent from the example of Afar and the Red Sea rifts that destruction is a complex process leading to transformation of the continental crust with the formation of an oceanic-type crust in the extreme case, such as the Red Sea and Aden Bay rifts and possibly the axial volcanic ridges of northern Afar. The above-mentioned data show that destruction took place as a result of: 1) shearing of the continental crust with the formation of rifts, gaping fractures and intrusion of basic dykes; 2) thinning of the granite-metamorphic layer due to its fragmentation, inclination of blocks and subsequent interaction between the plastic and brittle layers of the crust; and 3) saturation of the continental crust with material of basic composition, leading to deposition and sinking of the 'basalt' layer because of its possible transition to garnet-granulite facies, which follows from the concept of E.V. Artyushov [2].

Volcanism in Afar as an Indicator of Destruction of the Continental Crust

The Afar rift is the classical conjunction between two oceanic ('mini-oceanic') and continental rifts. In structure and development it shows features inherent for rifts. In this regard, the specific structures of Afar are the axial volcanic troughs participating in the form of active centres of expansion. Some researchers consider the axial volcanic ridges of Afar as continental analogues of mid-ridges of oceanic rifts. Similar analogies are based on the similarity of their morphology, deep-seated composition, nature of volcanism and dynamic circumstances of formation. Direct 'wedging' of structures of the Aden Bay axial graben into the Afar through Aden Bay confirms the proposed concept. The effect of this process is revealed in the change of petrochemical features of volcanism towards increased alkalinity of the oceanic tholeiites of Aden Bay as one approaches Afar.

Among the many concepts regarding the destructive processes of the continental crust, the one advocating successive development of the mantle diapir is better founded and conforms best to the evolution of volcanism and the geodynamic situation of formation of rift structures. V.E. Khain,

Yu. M. Pushcharovskii, E.E. Milanovskii, L.E. Levin and others impart an important role to this concept for the formation of marginal and internal seas of the Atlantic (passive) and Pacific (active) types.

Following the concept of the mantle diapir with the Red Sea rift as an example, development of the rifts in the region under study can be presented in the following manner (see Fig. 27). The mantle diapir apparently formed at great depth at an early stage (Oligocene-Miocene) and had the form of a large gentle sloping arch. Stretching in its axial part was accompanied by the appearance of a series of faults acting as paths for the penetration of alkali-basaltoid magmas, which in the course (146) of development of the diapir changed into weakly differentiated tholeiite-basaltoid magmas.

According to the concept of successive thinning of the continental crust, the evolution of alkali to tholeiitic volcanism took place in regions where the mantle diapir had developed. Available data show that the Ethiopian trap series, conformable with the early stage of development of the Afar rift, contain varieties transitional to the tholeiitic types together with alkali basalts and that this regularity is seldom broken. Further, this regularity is firmly proved from the example of the Kenyan rift zone [19].

Shearing which started in the middle Miocene subsequently became stronger as a result of horizontal spreading of the diapir. This led to the occurrence of an extensive basin, filled in by a thick predominantly evaporite-containing layer, which can be traced through seismic data. Formation of the rift valley was accompanied by the emplacement of undifferentiated basaltoid magma. On the one hand, the basaltic melt became consolidated in the form of a dyke at depth and on the other (along the coast of north-eastern Sudan), erupted to the surface forming a consolidated cover there. According to Colman and co-workers, a differentiated layered gabbro-massif formed at certain places, such as Tihama-Asir, in the eastern coast of the Red Sea under a favourable tectonic set-up.

The widespread major unconformity between the Miocene and Pliocene accounts for the upliftment and erosion of sediments and the same can be associated with recent upliftment of the mantle diapir and its subsequent interaction with granite-metamorphic layers which were earlier emplaced in the pre-existing large-scale trough of the Red Sea. The endogenetic process was active at this stage within a narrow zone of the axial trough and was accompanied by intensive injection of undifferentiated basalts in the form of vertical dykes and intrusives with fragmentation, faulting, and thinning of the continental crust. These processes continued until the granite-metamorphic layer was totally sheared and finally eroded.

Two generations of undifferentiated oceanic-type basalts of varying age and spatial separation were thus formed as a result of successive stages of development of the mantle diapir in the Red Sea rift. Herein, the younger stage of development of the diapir is characterised by lesser depth and correspondingly lesser 'dispersion' of basalt injections (dykes and intrusions). Its action is manifested in a narrow axial zone of graben and, possibly, completely 'transforms' dykes of earlier generations (Fig. 33). Under repeated pulsational development of the mantle diapir, this process is possibly capable of producing a series of mantle injections of different ages, with characteristic younging towards the centre. In any case, the existence of two generations of tholeiite-basalt magmatism (Miocene and Pliocene-Quaternary) for the Red Sea rift is undisputable and evidence of the principal possibility of such a process. According to N.A. Logachev, A.F. Grachev, A.E. Svyatlovskii and others, it conforms (147) well with the centripetal tendency of volcanism during the rifting regime distinctly manifested in the Afar, Baikal and Kenyan rifts.

According to F. Barbery and J.P. Varette, the model of development of the emerging mantle diapir essentially fits well with the volcanic evolution of axial ridges in Afar. The development of the latter took place in

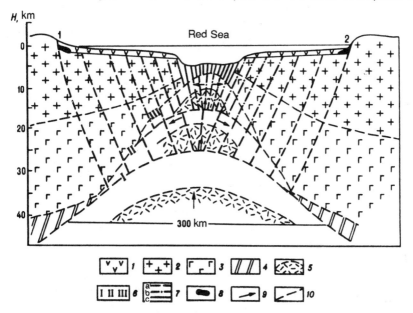

(147) Fig. 33. Development pattern of the mantle diapir in the Red Sea rift zone.

1—Neogene-Quaternary sediments (sketch without scale); 2 to 3—layers: 2—'granitic' 3—basaltic; 4—upper mantle; 5—roof of anomalous mantle; 6—provisional geoisotherms or heat-mass transfer fronts in the developmental stages of the mantle diapir: I—early (P_3-N_1^2), II—middle (N_1^3), III—late (N_2-Q); 7—basalt dykes (a—early, b—intermediate, c—late stage); 8—basalts: 1—Abu-Shagara of north-east Sudan, 2—Tihama-Asir, Saudi Arabia; 9—direction between tectonic shearing; 10—boundary between layers.

three stages. In the first stage (fissure eruptions or origin of expanding axes), eruptions of weakly differentiated olivine tholeiites, petrochemically close to basalts of mid-oceanic ridges (Fig. 34), took place. Tectonic and volcanic activity were manifested in narrow axial grabens and were accompanied by simultaneous formation of faults and intrusion of basalt dykes along both sides of the graben at distances restricted to less than 15 km. At this stage, the age of basalts decreases symmetrically from the margins to the axis of graben. This situation apparently depicts the characteristics of the deep-seated process. The second stage is marked by the formation of sheet basalts which led to accumulation of lava along the axial part of the original graben and gave rise to the elongated uplifts complicated by faults. Magmatic chambers, petrochemically enriched in iron of basaltoid differentiates, may become isolated at this stage. Central type volcanoes emerged in the third stage and gave rise to eruptions of highly differentiated lavas, such as trachytes, pantellerites and super-alkaline acid rocks—comendites. This was favoured by the stability of individualised magmatic chambers, formed as a result of lower tectonic activity. However, in the opinion of F. Barbery and J. Varette, magma generation took place at a considerable depth (at the mantle level), which is indicated by low $^{87}Sr/^{86}Sr$ ratio, equal to 0.7022.

F. Barbery and J. Varette compare the first and second stages with the mid-oceanic rifts and Hawaiian island volcanoes respectively. Petrochemical data show that volcanism of the axial ridges of Afar actually has definite similarity with volcanism of oceanic structures (see Fig. 34). As also in the case of development of the Hawaiian structure, tholeiitic volcanism changed to the alkali-basaltoid type up to eruption of alkali-trachytes, pantellerites and comendites along with their stabilisation and waning of tectonic activity. Development of the Afar volcanic ridges demonstrates in miniature, i.e., in highly reduced form, the evolution of oceanic structures in the continent during a short period (3–5 million years) from the stage of mid-oceanic ridges to the stage of relative consolidation of volcanic islands of the Hawaiian type.

(149)   Tectonic activity in Afar diminished with transition to more 'mature' stages of shield volcanoes. This tendency is better revealed in the last stage, when the central-type volcanoes become the dominant form of volcanic eruptions. Consequently, practically a full cycle of endogenetic development of rift zone from its origin to waning stages can be traced in the development of axial volcanic ridges of the Afar region.

In this regard, the interrelationship of the Red Sea rift and axial ridges of Afar is very interesting. It is considered that the Red Sea rift is an outcome of the more advanced stage of rift formation whereas the Afar rift represents stages proceeding to it. However, this concept is deceptive and is purely psychological, inasmuch as the Red Sea essentially rep-

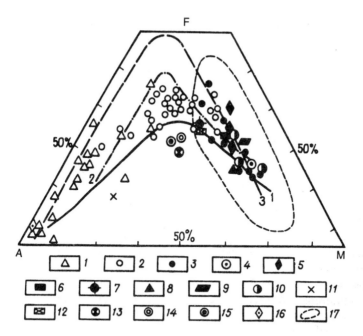

Fig. 34. AFM diagram of basalts in the Red Sea-Aden rift zones and oceans.

1 to 3—volcanic rocks in developmental stages of the Afar axial ridges: 1—central volcanoes, 2—sheet volcanoes, 3—fissure volcanoes; 4 to 9—tholeiites in: 4—Aden Bay, 5—axial trough of Red Sea, 6—Sheba and central Indian ridges, 7—Tair Island, Red Sea, 8—mid-Atlantic ridge, 9—Hawaiian Islands; 10—mid-Miocene basalts of Red Sea Coast; 11—mugearites of Hawaiian Islands; 12—Hawaiites; 13 to 15—alkaline basalts in: 13—eastern Pacific uplift, 14—olivine islands of the Indian Ocean, 15—olivine islands of Pacific Ocean; 16—comendites of Pashi Island; 17—field of oceanic basalts. Line of differentiation: 1—alkaline rocks, 2—tholeiites of the Hawaiian series, 3—Skargaard intrusions.

resents a 'mini-oceanic' structure. From the viewpoint of petrochemistry depicting the process of deep-seated evolution of the earth's crust, the volcanic axial ridges of Afar are represented by structures whose development in the 'continental variant' had already passed the axial trough stage of the Red Sea rift. It is not difficult to agree with this concept if we compare the most important petrochemical indicators for basalts of the axial trough of the Red Sea and axial volcanic ridges of Afar. The tholeiitic basalts of the Red Sea are comparable only to the volcanites of the initial stages of axial ridges.

Insofar as the relationship of axial volcanic ridges of Afar with mid-oceanic ridges is convincing and continental and oceanic rifts are considered different stages of a single process, i.e., rift formation, then it is logical to expect analogues of extra-axial areas of Afar in the ocean.

Zones of quiet magnetic field, and not oceanic plates (talus formation stage according to Yu.M. Pushcharovskii), may be such zones, as was initially proposed by the author. Quiet magnetic fields are characterised by specific compositional features, such as indistinctness of magnetic fields and interrupted nature of seismic waves, which are not inherent features of oceanic crust.

Extra-axial areas of Afar have a transitional type crust in their deep-seated structures which do not fully conform to the crust of the quiet magnetic zone at the given stage of rift formation. In the course of further development in rift formation, this crust could acquire features close to geophysical indicators for oceanic crust, as for example, in Aden Bay, continental margin of Southern Australia and other regions of the world. According to J. Cochran, the quiet magnetic zone in Aden Bay was formed at the cost of the continental means of diffused shearing and from a geophysical point of view is closely similar to the oceanic zone. On the basis of the continental margin of Southern Australia, M. Talvani and others established that the crust of the quiet magnetic zone is neither oceanic nor continental, but riftogenic, and that it played an important role in the evolution of a passive continental margin.

If rift formation in Afar was more intensely manifested in the formation stage of the main depression (N_1), then the newly formed transitional type crust during the period of formation of axial volcanic ridges in the Pleistocene-Quaternary period could have been traced over a large (150) area and, in the first place, at such places where the continental crust is presently thin. According to A.F. Grachev, in the case of further development of rift formation confined to the oceanic stage, one should have expected total detachment of continental blocks, such as Danakil Hills, Aisha Block, Tazura Bay, in the same way in which the continental relicts, such as the Seychelles and Sokotra islands and others, were formed. It is noteworthy that correlation of marginal parts of Afar with transitional type crust is still debatable.

Change in the composition of volcanites in the development of axial volcanic ridges of Afar serves as a good indicator for the dynamics of the endogenetic regime, manifested in the migration of magma generation level. At this point, the emerging and fading stages may be identified in the endogenetic regimes of rift zones. The emerging stage of development is revealed in the manifestation of alkali volcanism in the early stages of rift formation and in its change into weakly differentiated tholeiite in the late stage. This stage characterises an active destructional geodynamic situation (axial troughs of the Red Sea and Aden Bay). A change in tholeiitic volcanism by alkaline volcanism in the fading stage, on the contrary, shows an increase in the generation depth of magma and fading of tectonic activity. The initial stage of alkaline volcanism affected the axial volcanic ridges of Afar. The endogenetic regime of the fading stage was passive, apparently residual, and did not lead to active expansion.

Thus, with Afar volcanism and the entire Red Sea-Aden rift zone as examples, a relationship between the evolution of volcanism and geodynamic regime of the rift zone has been established. In the early stages of the volcanic cycle, the sources of endogenetic stimulation being situated at great depth, they produced melts of more deep-seated and more alkaline magma. Subsequently, with strengthening of the endogenetic process, the degree of partial melting increased sharply, the magma generation level rose and weakly differentiated basalts were formed.

In the context of the above statements, the problem regarding the direction of volcanism (magmatism) during the rifting regime should be touched upon once again, inasmuch as Yu.G. Gatinskii and others consider that volcanism during rifting starts with the tholeiitic type. According to this concept, the directional variation in the composition of magmatism is linked with deeper penetration of fissures and subsequent draining of more deep-seated alkaline magma. The inaccuracy of this concept was clearly proved by A.F. Grachev in his book *Riftovye zony zemli* [Rift Zones of the Earth]. In 1980, A.V. Razvalyaev established that volcanism during rift formation evolves according to the development of the endogenetic regime and, consequently, with the degree of advancement of the mantle diapir. In the full tectonomagmatic cycle, volcanism evolves from the alkaline to tholeiitic, and once again reverts back to the alkaline type. The emerging and fading stages of endogenetic development of rift zones are distinguished following the above cycles. These concepts have been confirmed in the recent publications of A.I. Almuhamedov, G.L. Kashintsev and V.V. Matveenkov and are very important for understanding rift formation [1].

(151) The specificities in the evolution of the Afar crust extend much beyond the limits of regional significance. They represent particular interest in the light of problems of development of intra-continental and marginal seas [17]. The characteristics of destruction of the earth's crust established here may be extended for the development of internal seas, e.g., Tirreng, Aegean, Black Sea etc. and also for marginal seas located in the zones of active continental margins, such as Okhotsk, Japan, Bering and south-east Asian seas. The mechanism of destruction of the continental crust is more often drawn for explaining their nature and, in the opinion of V.E. Khain, L.E. Levin, V.M. Kovylin and others, they are close to the Afar and Red Sea in their main features. The role of destructive tectogenesis in the evolution of the earth's crust increases sharply if it is considered that similar processes are at the base of development of passive continental boundaries.

The specificities in the composition and development of Afar help in understanding the nature of destructive structures in the geologic past, especially the rift structures of Western Siberia and, possibly, palaeorifts (aulacogen) of Western Europe and Russian plates. In contrast to linear or fractured rifts, in which endogenetic energy is concentrated along narrow zones, these regions are characterised by extensive lateral shearing, shallow depth of occurrence of anomalous mantle, presence of differently oriented, often paired and contiguous axes (centres) of expansion and occurrence of endogenetic energy in a dispersed manner. According to V.V. Belousov, a similar dispersed endogenetic regime is created in those cases when the crust is still inadequately consolidated even at the time of rift formation.

The Western Siberian grabens, distributed over a large territory, have the potentiality to be closer in formation to an oceanic-type crust in geodynamic status and endogenetic regime. In this respect, they have been more correctly characterised, as never before, by V.V. Belousov [3] who named Western Siberia an 'unsuccessful' ocean. In the case of Afar, whose mature continental crust was formed in the Precambrian and the dispersed regime emerged as a result of fracturing of the lithosphere at the trijunction of three rifts, a mozaic combination of tensional vectors created an unstable dynamic state of the lithosphere, which favoured a dispersed appearance of the endogenetic process. In distinction to Western Siberia, the ocean was 'successful' in Afar, particularly in the northern part (Danakil depression), but was manifested as a 'dryland' variant.

Depth Aspect in Destruction of the Continental Crust in Regions of Active Continental Rift Formation

Conversion of the continental crust into an oceanic one in the course of time and space is manifested as the result of a dynamic process in the upper mantle and crust. Knowledge of this process is very important, insofar as it brings one closer to an understanding of one side (destructive) (152) of the interrelated constructive-destructive process of tectogenesis. This process may be traced in the complex geologic history of regions of active continental rift formation of the Red Sea and Afar rift type as a series of stages with varying degree of development in the mantle diapir. The appearance of newly formed oceanic-type crust in place of continental crust marks its final stage of development.

One interesting aspect of the deep-seated evolution of the continental crust under rift formation is its transformation into oceanic crust. Did this process take place through rupture of the continental crust along a singular vertical plane with subsequent separation of diverging blocks (plates) or did some more complex transformation occur here? Geophysical and petrologic data show that this process was prolonged and proceeded dissimilarly in different levels of the lithosphere. Rift formation associated with intrusion of thinned mantle material into the lithosphere caused a complex process of material transformation in the crust. One expression of this process is the appearance of a layer (intermediate according to S. Müller) with a velocity of 6.3—6.4 km/sec in the sialic crust.

Concomitant with progress of the deep-seated process (mantle diapirism) during early stages of rift formation a low, viscous, plastic layer formed in the lower horizons of the sialic crust in which plastic deformation predominates owing to enhanced temperature. The viscosity of rocks sharply decreased under such conditions and their mobility increased [2]. This and particularly the underlying 'basalt' layer became saturated with magmatic injections of mantle origin (basalt dykes in the axial trough of Red Sea rift, Afar). Further, a 'collar' formed in the plastic sialic layer, which progressively thinned and was severed in more developed rifts, and the underlying layer became more saturated with mantle material and converted into a 'pseudo-oceanic' crust, which is the third oceanic layer.

A heated sialic layer is amenable to plastic deformation under tension; this fact has been established in many active continental rifts at different depths depending on the stage of their development. An intermediate layer (6.0—6.4 km/sec) is detected in the upper parts of the crust in Afar (including the Danakil horst), Jibuti and in the northern part of the Kenyan rift; this layer overlies a layer which registers 6.7—6.9 km/sec velocity for the longitudinal waves. In central Afar this layer occurs at a depth of 3—5 km and has the same thickness. It tapers out in the proximity of 'oceanic' structures of the Red Sea and Aden Bay. The presence of a layer 3—4 km thick with 6.0—6.1 km/sec wave velocity in the northern Afar shows that although the upper crust under Afar distinctly differs from the 'normal' oceanic crust, it is nonetheless close to the latter. Numerous injections of basalt dykes, as noted earlier, so greatly transformed the initial continental crust that it is now 'intermediate' between continental and oceanic crusts.

(153)   The mantle diapiric process in the Red Sea (axial trough) and Aden Bay, which represent development of intercontinental rifts, led to the total disappearance of the intermediate layer. Results of seismologic investigations by R.V. Girdler record that a seismic layer with 6.6—8.0 km/sec velocity, characteristic for the second oceanic layer, is exposed in the axial trough. Dragging of rocks at the bottom of the axial trough and geologic-geophysical investigations conducted by the Oceanology Institute of the Academy of Sciences, USSR in the region of 18° N lat., including direct survey of the bottom by means of the automatic underwater apparatus 'Paisis', confirmed the presence of a second oceanic layer covered by thin sediments [19].

Relicts of the intermediate layer in the Red Sea rift are possibly preserved in its shelf zone and overlapped by thick (up to 6 km) predominantly evaporite rocks of Miocene age [12]. Some geophysical data favour such a postulate. Thus, seismic profiles revealed an acoustic basement with velocities equal to 5.91, 6.37 and 6.09 km/sec respectively, situated west of the axial trough (Nos. 175–176, north-east of Dahlak archipelago; No. 177, 18° N lat. in the southern and No. 182, 24° N lat. in the northern part of the Red Sea). Such an acoustic basement could be identified with the sialic layer saturated with basalt dykes.

The progressive development process of crustal thinning during rift formation is interrelated with volcanism. It is suggested that a moment arises in the process of thinning of the continental crust together with formation of a 'collar' wherein further thinning of the crust associated with a thinned mantle leads to severance of the 'collar'. At this stage, the permeability of the crust increases. It is logical to presume that volcanism increases sharply mainly from this moment. Apparently, the Kenyan, Ethiopian and Afar rifts belong to this stage or stages closer to such development. It follows from geophysical data that the thickness of the intermediate sialic layer in these rifts amounts to 3–5 km approximately, although the depth of occurrence varies.

The geophysical meaning in the evolution of volcanism lies in the variations of the deep-seated constitution of the crust. Emplacement of weakly differentiated basalt and diabase dykes and fissure-controlled eruptions indicates formation of an intermediate layer (6.3–6.4 km/sec) in them. Consequently, the massive intrusion of basalt dykes may be regarded as an indicator of specific changes in the deep-seated constitution of the continental crust in rift-forming zones.

Thus it follows from a short review of the deep-seated constitution of the Red Sea and Afar rifts, that the deep-seated aspect in the evolution of the continental crust during rift formation consists in the appearance of a layer of 6.3–6.4 km/sec velocity in the lower part of the sialic layer and in its progressive thinning with the advent of destruction of the continental crust 'collar' in the intermediate stage. In the evolutionary process of the crust, the intermediate sialic layer tends to thinning and total rupture, whereas the lower layer tends to thickening and 'basification'. The progressively developing rupturing process of the continental crust in more developed rift structures (Red Sea) initially led to the formation of a transitional crust in the axial trough and, with further thinning, to its total severance in the axial trough and formation of a crust similar in geophyical indicators to the oceanic type, but still apparently holding the continental blocks. The presence of components of continental crust in the axial trough was recorded in Chapter 3, through orientation of a fault network both in water areas, including the axial trough, as well as in the

(154)

continent. At this stage, the axial trough apparently does not represent a gaping fracture along the entire depth of the lithosphere.

The following conclusions may be drawn on the basis of the foregoing:

1. Evolution of the earth's crust in regions of active continental rift formation is manifested in the destruction of continental crust and revival of oceanic crust.

2. Destruction takes place by fragmentation and shearing of the continental crust, surfacial and apparently subcrustal erosion and intrusion of basic dykes that lead to a change in geophysical parameters as they approach towards the characteristics of a 'basalt' layer.

3. Features of volcanism and dynamics of development of suture zones of the Red Sea, Aden Bay and East African rifts help us to delineate an analogy of the oceanic rifts in the axial volcanic ridges with their continental variant.

4. The example of Afar helps to identify the early stages of development of mozaic structure, characterising the complex geodynamic set-up, appearance of dispersed endogenetic regime and emergence of one of the specific forms of destruction of the continental crust.

5. The deep-seated aspect of evolution of the continental crust during rift formation is manifested in mantle diapirism. the transition process from continental rifts to the oceanic types is accompanied by formation of layers conforming to the second layer of the oceanic crust (Red Sea) in geophysical and petrologic parameters. This 'pseudo-oceanic' layer or, as it were, the 'residual' layer from the continental crust, is replaced by a 'natural' oceanic crust (Aden rift) in the course of further development of rift formation.

Thus, evolution of the earth's crust in regions of active continental rift formation leads to its destruction and replacement by a newly formed crust of oceanic type. The destruction represents a complex process of varying physico-chemical properties of the crust, caused by intrusion of asthenospheric diapirs into the lithosphere, facilitating its heating and shearing. At this stage, shearing is regarded as a result of intrusion of asthenospheric diapirs. A similar concept assigns an active role to interactive processes of the asthenosphere and lithosphere in tectogenesis [20] and sharply lowers the significance of passive intrusion of the asthenosphere into the lithosphere during its splitting due to forces applied from outside, as has been admitted from the position of the new global tectonic process. At the same time, the author does not exclude the possible effect of global zones of activisation in tectogenesis, identified by Yu.G. Leonov [17], on the mantle diapiric process, which acted as an accelerating factor for rift formation.

(155)

6

Endogenetic Regimes of Continental Rift Formation and Its Pattern of Emergence

Endogenetic Regimes of Continental Rift Zones and Principles of Their Distinction

(155) The necessity of establishing the patterns of interrelationship of rift formation with the preceding developmental stages becomes more evident in the study of continental rift formation and problems concerning its predetermination and inheritance are now introduced. The publications of E.E. Milanovskii are of great significance in revealing these characteristics; they distinguish 'maturing' of the prerift stages of Cenzoic rifts, which were manifested tens and even hundreds of million years before their formation. Similarly important are the publications of V.V. Belousov, N.A. Florensov, N.A. Logachev, N.A. Bozhko, A.F. Grachev, E.A. Dolginov, V.P. Ponikarov and A.V. Razvalyaev and others, covering deep-seated thermal activity and evolution of the mantle in regions of continental rift formation. These problems have been dealt with in general by geologists outside the USSR and are contained in the publications of F. Dixie, R.B. McConnell, V. Fybe, O. Leonardos and others.

The relationship of continental rift zones with basement structures ('frame') and their anisotropy is well known. It has been established that the closer relationship of continental rift formation with preceding history consists in its confinement to regions with Precambrian structure, which preserved much tectonic activity at the time of rift formation and, consequently, a more intensive thermal regime also [8, 19, 27]. This includes extrageosynclinal regions which were subjected to repeated tectonomagmatic activisation (e.g., the Mozambique and Libyan-Nigerian belts in the late Riphean-early Cambrian period; Grenville and Dalsland belt in the middle Riphean period).

N.A. Bozhko, E.A. Dolginov, V.M. Moralev, V.P. Ponikarpov developed the ideas of V.E. Khain on the uniqueness of the late Proterozoic tectonomagmatic activisation belts and their predisposition to rift formation and marginal location in the continent-ocean system. The uniqueness of these belts is manifested in the presence of rocks of granulite facies (156) of metamorphism, i.e., charnockite and anorthesite massifs among them. In their structural set-up the belts are regarded as weak zones representing places of separation of continents and emplacement of newly formed oceanic basins. N.A. Bozhko considers that continental rift formation preferentially takes place in the basement and its main features are represented by mafic granulite-basite composition and deep-seated tectonothermal transformation.

Thus it is generally known what structural and geologic conditions are more predisposed to rift formation (tectonothermally activised late Proterozoic belts, basitic profile of the substratum and its late Proterozoic age and others) and what situations are not typical. What is not yet known is why the rifts, originating purportedly under similar geologic-structural conditions, attained different stages of individual development. For example, the degree of 'maturity' of the Red Sea and Baikal rifts differs, albeit they were laid on the Baikal basement which underwent repeated extrageosynclinal tectonomagmatic activisation. Although theoretically alkali and alkakli-ultrabasic magmatism precede rift formation and are an inseparable part of its preparatory stage, even then it is still not clear why rift formation in all regions was not preceded by the development of alkali and alkali-ultrabasic magmatism. Thus Mozambique type continental rifting of this marginal continental belt did not develop in all similar belts and is also not identical everywhere. It should further be noted that although the Red Sea rift is situated in the northern continuation of the Mozambique belt, it cannot be classified as a continental-marginal type. Finally, all rift systems can hardly be related to Mozambique type belts. Consequently, the features mentioned above, correct though they be, still do not fulfil the geo-historical prerequisites of continental rift formation.

It follows from the aforesaid that the degree of rift formation differs significantly under a similar prehistory. Apparently, the constitution and development of rift zones, as well as their relationship with the substratum, is much more complex than the known interrelationships. In addition to general features of constitution and development, the rift zones are characterised by considerably large differences, the nature of which is not yet known in many cases. It is not yet known what processes and what combination of conditions, other than the observed, are necessary for rift formation; in other words, what optimal assemblage of geologic 'situations' or endogenetic regimes are favourable for rift formation.

A study of the regions of contemporary active continental rift formation and its interrelationships with prehistory is of major significance for understanding rift-formation processes. Growing interest in the problems of rift formation favoured the emergence of publications dealing with problems of riftogenetic magmatism and distinction of palaeorift regimes. Among them, the study by V.N. Moskaleva [23], devoted to magmatic formations as indicators of rift-forming systems, deserves special attention. She noted that the identification of ancient riftogenic systems is, first of all, possible on the basis of the products of magmatism formed; she also stressed that a study of the chronological and lateral order of formations in rift-forming zones is of major significance for reconstructing palaeorifts. Insofar as magmatism, an important indicator of rift formation, is manifested not only in grabens (rifts), but also extends much beyond their boundaries, its study is difficult in many rift systems. Hence, identification of the full chronological and lateral order of magmatic formations poses specific problems and is not possible in all rift-forming regions.

(157)

From the above-mentioned point of view, problems of rifting in the Red Sea-Aden rift zone serve as a unique polygon for revealing the geological history of rift formation insofar as extrageosynclinal activisation magmatism, which is the main indicator for endogenetic regimes, is extensively developed in its prerift development stage. The volcanic processes are widely manifested here, directly preceding and accompanying graben formation up to the Recent period, and Cenozoic rifts of the Red Sea and Aden Bay attained regeneration of the oceanic crust during the stages of their development. The Red Sea and other continental rift zones showed, as examples, the prerift endogenetic regimes (situations which preceded Cenozoic rift formation).

Development of the Red Sea rift zones during the postgeosynclinal stage took place after the consolidation of the late Proterozoic Red Sea fold belt. This is characterised by prolonged tectonomagmatic activisation, encompassing almost the entire Palaeozoic and Mesozoic. Activisation is revealed through the appearance of new and restoration of old dislocations and magmatism. Alkali-gabbroid, alkali-olivine basalt (trachybasalt), tholeiite-basalt, nepheline-alkali-syenite and alkali-granitoid complexes were formed in the prerift stage of development of the Red Sea zone (Fig. 35). The alkali-ultrabasic complexes are provisionally delineated. The prerift deep-seated basaltoid (alkaline and tholeiitic) magmatism developed during the following age limits: 680–450, 290, 250, 185, 120 and 80 million years. The alkali-granitoid magma, occurring alternately or synchronously with basaltoid magmatism and occupying crustal level, is also characterised by multiplicity and occurred at 570–450, 185, 120 and 50 million years age [as per age determination conducted outside the USSR*].

It follows from the aforesaid that a prolonged, multifaceted post-geosynclinal tectonomagmatic activisation of the 'penetrative' type preceded

---

*Addition ours—General Editor of English translation.

(158) Fig. 35. Chronological order of magmatic complexes in the Red Sea rift zone.

Formations: 1—alkali-granitoid; 2—alkali-gabbroid; 3—alkali and nepheline syenites; 4—differentiated intrusions of gabbro-norite-anorthosite composition ('layered' gabbro); 5—alkali-ultrabasic rocks (?); 6—alkali olivine-basalt; 7—tholeiitic basalts; 8—andesite-liparite.

Cenozoic rift formation in the Red Sea. Epochs of more active magmatism were recorded in the Vendian, Cambrian, Ordovician, Silurian, Devonian, Permian, Jurassic, Cretaceous, and Palaeogene. The depth of the level of magma generation changed periodically in the course of the above time interval. The magmatic process took place, as it were, in pulsations. The ancient magmatic activisation epochs were more prolonged and the younger epochs were of short duration and more frequent. The characteristic features of magmatism in the Red Sea rift zone during the (158) prerift stage are marked by multiple alternations of alkali-gabbroid, i.e., alkali-olivine-basaltoid, tholeiite-basaltoid and nepheline-alkali-syenite complexes, since products of deep-seated mantle-based magmas and alkali-granitoid complexes from magma at the crustal generation level. The concluding and inherently rifting stage in the Cenozoic of the Red Sea-Aden rift zone is characterised by the development of alkali-olivine-basaltoid and tholeiite-basaltoid (trap) complexes. Cyclicity is distinctly manifested in the evolution of volcanism in the inherently rifting stage of the Red Sea-Aden rift zone. Two major impulses (cycles) are distinguished, early (Eocene-Miocene) and late (Pliocene-Recent).

Products of the earliest sources of volcanism during the rifting stage are widespread in Ethiopia, Yemen Arab Republic, Sudan, Saudi Arabia and in more remote regions from the Red Sea, i.e., Aden rift zones (Libya and Syria). These have been better studied in Ethiopia, where they constitute an extensive plateau and are known by the term 'trap series'. Following the data of P. Brotz, E. Pikirillo and others, the trap series is petrographically varied and its formation reflects complex evolution during the prerift stage. It has been established that the series consists predominantly of weakly differentiated Ashangi basalts (60–45 million years), intermediate (or transitional) in composition between the tholeiitic and alkali-basalts of the Aiba complex (34–28 million years), bimodal volcanites (basalts and rhyolites) of the Alaji complex (32–26 million years) and basalts with a subordinate quantity of Termaber phonolite complex (25–13 million years).

Petrochemical data show (Fig. 36) that the oldest Ashangi complex comprises sheet alkali basalts in the lower part and weakly differentiated (159) tholeiitic basalts in the upper part. Its formation was concluded by the rift-forming phase, upliftment of the territory and peneplaination (the Ashangi peneplain). The new impulse of volcanism was accompanied by the Aiba basalt eruptions, which are transitional in composition, but significant for tholeiitic varieties. Subsequent eruptions became more alkaline (Alaji rhyolites and basalts). The volcanic cycle of the Ethiopian plateau was concluded by significantly alkaline Termaber basaltic eruptions, followed by the formation of the Ethiopian rift in the Miocene.

The following volcanic cycle is distinctly manifested in the Afar depression and has already been detailed in Chapter 5 in relation to the evolution of volcanism and its role in the destruction process of the continental crust in active continental rift-forming regions.

A study of Cenozoic volcanism in the African-Arabian rift belt helped in reaching a conclusion regarding the dependence of its composition on the geodynamic regime. Thus, volcanism started with rocks of higher alkalinity in the earliest stages of development of rifts. This stage is more distinctly manifested in the Kenyan rift through the appearance of nephelinite-phonolite associations with carbonatites. The magma generation depth decreased side by side with advancement of the mantle diapir, and alkali-olivine-basalts and their differentiation products (alkali-olivine-basalt, basalt-rhyolite-pantellerite and basalt-phonolite-pantellerite associations) were formed. This rift-forming stage is distinctly manifested in the Kenyan rift zone, the Ethiopian rift and the Afar depression. Formation of the central graben coincides with it. Volcanism concluded with the appearance of comendites and pantellerites (Afar) in more developed rifts. Herein, the destruction process of the continental crust led to its substantial thinning; however, D.K. Bailey and others considered that it did not lead to severance. In more developed (mature) rifts (Red Sea, Aden), where the mantle diapirism process led to complete destruction of the continental crust and its replacement by regenerated oceanic type crust, the typical oceanic tholeiites were already formed (see Fig. 33). Thus, the composition of volcanism in rift zones depends upon the degree of development of the mantle diapir or tectonothermal systems.

Volcanism undergoes evolution in the full tectonomagmatic cycle, from the alkaline to less differentiated types and again to the alkaline type. However, associations of incomplete or shortened cycles are more often found in reality; their individual members are either absent or are highly shortened. For example, the early alkali-basalt stage is practically absent in the axial volcanic ridges of Afar, which were active centres for destruction of the continental crust. This stage is significantly reduced in the Ethiopian plateau, also formed under high tectonic activity, and is represented by the Ashangi alkali basalts. The stage is more fully represented in the Kenyan rift [19]. Directed evolutionary development of the volcanic cycle may be complicated or even interrupted. Thus, for example, the Ashangi volcanism in the Ethiopian plateau was interrupted by (161) the rift-forming stage and peneplaination (the Ashangi peneplain is 45–34 million years old). Consequently, pulsational change also takes place in magma generation depth in the course of development of the endogenetic regime of an essentially rifting stage. Such a circumstance shows the similarity and succession in development of zones during continental rift formation. Besides, conclusion of the pulsational character of appearance of volcanism during continental rift formation speaks in favour of further development of concepts on the pulsational nature of tectogenesis in general.

The characteristics of volcanism in the Red Sea-Aden rift zone are interesting in the aspect of problems concerning destruction of continents. It is known that a specific feature of this process during the rift-forming regime of tectogenesis is the existence of a geodynamic situation of ex-

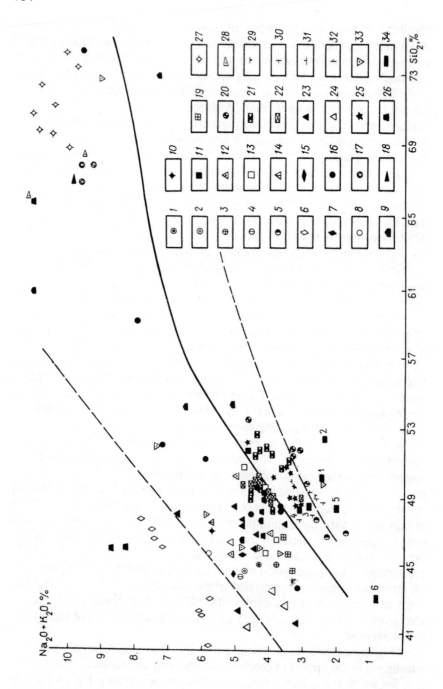

tensive stretching (shearing). Among many prevailing viewpoints on the destruction process of the continental crust, the one on manifestation of successive development of the mantle diapir conforms best with the evolution of volcanism and the geodynamic situation of formation of rift zones. From the position of mantle diapirism, a different degree of differentiation of magma is considered to be dependent upon the degree of 'advancement' or development of the emerging (rising) mantle diapir. The characteristic spatial distribution of magmatic sources in the rifting stage of development of the Red Sea-Aden rift zone, different degrees of differentiation of its products, non-uniformity and discreteness in their distribution help in proposing the concept of multiplicity of mantle diapirs situated in a different stage of development ('advancement') and representing itself as projections (asthenoliths) of the thinned mantle (Fig. 37).

The more advanced (developed) process of mantle diapirism in such rifts as the Red Sea, Aden and, possibly, Afar (axial volcanic ridges), led to destruction of the continental crust—thinning of the granite-metamorphic layer and its subsequent complete severance in the axial troughs of the Red Sea and Aden Bay with the formation of oceanic-type crust. Structurally, it was manifested in the appearance of rift-generated faults and the grabens accompanying them. In other regions extending to the Red Sea-Aden rift zone, such as Tibesti, Haruz (Libya), Bayuda, Marra (Sudan), the Syrian volcanic plateau, Saudi Arabia and Yemen Arab Republic, where the mantle diapirism process is manifested in rudimentary or less advanced (reduced) form, only embryonic development of arched uplift-
(162) ments took place, directly preceding rift formation. The question arises

---

(160) Fig. 36. Correlation between alkalis and silica contents in rift zone volcanic rocks of the northern part of the African-Arabian belt.

1 to 4—basalts of Syria: 1—middle Miocene, 2—Pliocene, 3—Quaternary, 4—contemporary (Recent); 5 to 8—basalts of Sudan: 5—middle Miocene, 6—Bayud desert, 7—Nile valley, 8—Khartoum region; 9 to 13—basalts of Libya: 9—Tibesti, 10—Haruz, 11—Chirian, 12—southern Tibesti, 13—Chad; 14 to 15—basalts of Yemen People's Republic: 14—trap series, 15—Hamdan; 16 to 18—volcanites of Afar: 16—stratoidal type, 17—pantelleritic ignimbrites, 18—alkali rhyolites; 19 to 24—volcanic rocks of central Ethiopia: 19—alkali basalts of Ashangi, 20—tholeiites of Ashangi, 21—Aiba, 22—Alazi, 23—Termaber, 24—Termaber basanites; 25 to 28—volcanic rocks of south-eastern Ethiopia: 25—transitional basalts of thoeliitic type, 26—transitional basalts of alkaline type, 27—pantelleritic obsidians, 28—rhyolites; 29 to 33—basalts of Red Sea, Aden Bay and adjacent oceans: 29—Aden Bay, 30—axial trough of Red Sea, 31—Tair Island, 32—oceanic tholeiite, 33—alkali olivine-basalts of Red Sea islands; 34—traps (1—Deccan, 2—Karroo, 3,4—Siberia, 5—Comatiit of South Africa, 6—pyrolite (according to A. Ringwood)). Solid line divides the field of subalkaline and alkaline basalts (according to G. McDonald and T. Katssura); broken line marks the boundaries of transitional type basalt fields.

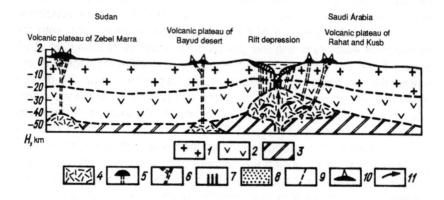

(162) Fig. 37. Pattern of deep-seated structure of the Red Sea rift zone.

1 to 2—layers: 1—metamorphic granite type, 2—basaltic; 3—upper mantle; 4—anomalous mantle: 5—sheet volcanics; 6—magma feeder canals; 7—basic and ultrabasic dykes—newly formed oceanic type crust; 8—sediment filling of grabens (approximate); 9—faults; 10—volcanoes; 11—assumed stress directions.

whether these regions have any relationship with rift formation if the latter are not structurally revealed here in the form of grabens?

A characteristic feature of these Cenozoic volcanic phenomena is the absence of any distinct relationship with taphrogenic structures. Also, they are controlled by such structures, as, for example, Tripoli-Tibesti, and are confined to the dome-shaped most recent upliftment [37]. Basalts of the Arabian Peninsula and northern Africa conform to alkali-olivine-basalts in composition with some petrochemical variations. It is important to know the reasons that gave rise to the inhomogeneous distribution of volcanic areas in North Africa and Arabia. Based on the petrochemical similarity of basalts the author put forward the assumption that all such areas are related to a singular tectonomagmatic process, caused by activisation of the mantle, its heating and thinning [27]. The petrochemical variations of volcanites are explained by the presence of mantle diapirs or asthenospheric projections of different stages of development under each volcanic area. This assumption has been confirmed on the basis of gravimetric investigation results and it has been shown that the volcanic areas of Darfur, Tibesti, Hoggara type and others are related to the isometric negative Bouguer anomaly ($5 \cdot 10^{-4}$ m/sec^2), which may be treated as the appearance of vast lenses of thinned mantle [37] analogous to geophysically better studied Kenyan and Ethiopian dome upliftments. Analogous volcanic areas having no direct relation to tectonic structures are characteristic also for other regions, for example, the Baikal-Mongolian region, where their nature is similarly explained by the evolution of anomalous mantle [7].

A.F. Grachev and others showed that the appearance of volcanism over an extensive territory in the Baikal-Mongolian region evidencing no structural control, is also associated with the formation and upliftment of the asthenospheric layer. The correlation in petrochemical relationship of the basalts in the Mongolian People's Republic and Pribaikal helps to include them under a single endogenetic regime. Only in the Mongolian People's Republic is this regime characterised by a situation innate for the early rift formation stage in the Pribaikal. On this basis, A.F. Grachev distinguished a prerift regime or prerift stage. Accordingly, a prerift stage was established in Northeast Africa and Arabia as well as in the Baikal-Mongolian region. This stage preceded the stage when gentle but already defined linear depressions were formed. In distinction to the prerift stage distinguished by N.A. Florence and N.A. Logachev and distinctly manifested in the Red Sea rift during the Oligocene-Miocene, this stage directly precedes rift formation and corresponds to the formation period of volcanism, spread over extensive territories with no distinctly revealed structural control. As regards the depth of formation, this stage corresponds to formation of the asthenospheric layer and its inhomogeneous upliftment. Delineation of this stage is important for palaeogeodynamic reconstruction. The author also delineated a prerift and essentially rift stage for the rifting epoch which, in the concept of N.A. Florence and N.A. Logachev, later more successfully termed by them as the early and late stage, helped to preserve the terminology 'pre-rift stage'.

Thus the nature of the endogenetic regime in the arched volcanic regions and structurally controlled rifts is the same. The geodynamic environment for either is characterised by stretching. The arched volcanic regions represent an early stage of 'developed' rifts. Rift formation of analogous regions may be termed as embryonic or in the initial stage.

The formation of rifts from the viewpoint of development of the endogenetic system and stages of mantle diapirism using Red Sea rift zone as an example, may be represented in the following form (Fig. 38). Local (source based) zones of thinned mantle were formed in the prerift epoch which produced small volumes of magma and were injected in the form of intrusives of the central type. Multistage sources of magma at different times and depths existed in this epoch and formed the polychronous Arabian-Nubian alkaline provinces. The tectonic situation, characterised by weak stresses due to stretching, occurred in reactivised old fault zones and led to increased penetrability of the latter. Along with progress of the prerift endogenetic regime, the individual sources united into a considerably larger volume of accumulated melts (spatial stimulation of the mantle) and the asthenospheric projection (mantle diapir) or a single heat-mass transfer front was formed. A similar endogenetic situation conforms to the essentially rifting epoch in its early prerift stage. The mantle diapir possibly occurred at great depth during the Oligocene-Miocene and was represented by a gentle arch. Stretching in its arched part accounted for (164) the appearance of a series of new and reactivised ancient faults that served as conduits for penetration of alkali-basalt magma, which subsequently became transformed into a weakly differentiated tholeiite-basalt magma with the development of the diapir.

In conclusion, let us review a few general characteristics arising from the analysis of magmatism in the Red Sea-Aden rift zone. Firstly, high in-(165) consistency in the temperature of the deep interior at the mantle level accounts for transition into endogenetic regimes and their pulsational manifestation. The geodynamic situation in the mantle was characterised by repeated rythmic migration of magma generation zones in a vertical direction. Endogenetic stimulation of the mantle was localised along specific zones which trace the earliest sources of its stimulation. The linearity of prerift stimulation zones of the mantle helps us to segregate them as long-existing, activisation 'cores' or 'rift-forming cores'.

The chronological series of magmatic complexes in the prerift epoch reflects an evolutionary succession of endogenetic regimes of different depth and gives an idea about their dynamic characteristics, as well as about the thermal state of the crust and the mantle. The mantle was more heated during basalt magmatism and was capable of melting basic magmas. The formation of alkali-granitoid rock complexes per se attests to the uprising thermal front and formation of 'crustal' palingenetic magmas. Accordingly, epochs of elevated and lowered heating of the mantle may be distinguished during the prerift evolution of the Red Sea rift zone, i.e., periods of thermal stimulation and calmness which, according to V.V. Belousov, correspond to stimulated and quiescent endogenetic regimes.

It should be noted that a stimulated endogenetic regime is characteristic of the prerift as well as essentially rifting epochs of the Cenozoic rift genesis and is included by some researchers under the basaltoid activisation type. A conclusion may be drawn from the aforesaid that an analogous endogenetic regime is inherent in the prerift and rifting epochs, but prolonged and consistent manifestation of one-type endogenetic regimes attests to the inheritance of their development in regions of epiplatform continental rift genesis. The endogenetic regime of an essentially rifting epoch serves, as it were, as the logical conclusion of the prerift epoch, corresponding to the final stage of directed evolution of endogenetic regimes in continental rift genesis.

Are not these evolutionary features of endogenetic regimes characteristic only for the Red Sea rift, i.e., for rifts which proceeded far in the

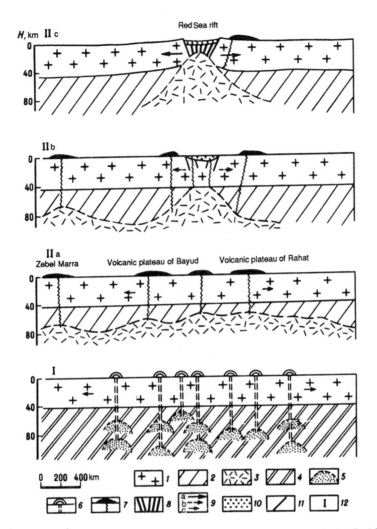

(164) Fig. 38. Development pattern of deep-seated structures and magmatism in the Red Sea rift zone.

1—earth's crust; 2—lithospheric part of mantle; 3—asthenosphere; 4—upper mantle; 5—sources of magma generation; 6—alkaline ring intrusions and their magma feeder canals; 7—volcanic plateaus and their feeder canals; 8—basic and ultrabasic dykes; 9—vectors (a—strong, b—moderate, c—weak shear); 10—sediments; 11—faults; 12—epochs and stages of evolution: I—prerift, Vendian-Eocene, II—rift formation and its stages (a—prerift, Eocene (?)—Oligocene, b—early, Oligocene-Miocene, c—late Pliocene-Quaternary).

destruction process of the continental crust? To answer this question, let us compare the endogenetic regimes of the prerift epoch of the Red Sea rift with those of the epiplatform rift zones, namely Baikal and Dahomey-Nigeria, which conform to different stages of development of Cenozoic rift formation (Fig. 39).

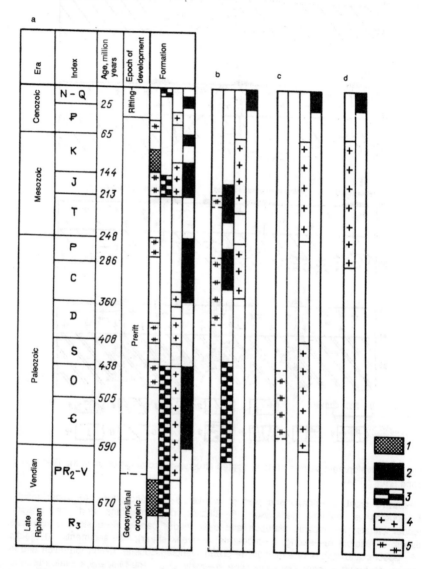

(166) Fig. 39 Chronologic formational series of (a) Red Sea, (b) Baikal, (c) Hubsugul rifts and (d) Dahomey-Nigerial rift zone.

Formations: 1—alkali-ultrabasic (?); 2—alkali-gabbroid (alkali olivine-basalt); 3—tholeiite-basalt; 4—alkali-granitoid; 5—nepheline-alkali-syenite.

The extrageosynclinal activised magmatism in the prerift epoch of the Baikal rift zone was manifested in the middle and late Palaeozoic-Mesozoic and its basement completed the geosynclinal orogenic development in the Vendian-early Cambrian. It is represented by three groups of complexes: Alkali-ultrabasic carbonatite, alkali-gabbroid and nepheline syenite-alkali granitoid. A.A. Konev delineated the following main epochs of alkali magmatism from radiometric age data: 650—550, 520—400, 350—300, 220—190, 140—110 and 50—1 million years [14].

The alkaline rocks, distributed unevenly over an area, are grouped in linear extended belts (provinces): Northern Baikal, Vitim, eastern Sayan, Prisayan and Jidinsk. The more extended northern Baikal alkaline belt is represented by ring intrusives of the Synnyrsk complex; its zone of development coincides with the Baikal rift axis and its extension can be traced in northern Pribaikal up to the middle course of the Mama river for a distance of less than 400 km. Such alkaline rock massifs as the Synnyrsk and Akitsk belong to this belt. According to the data of A.A. Andreev, the age of the Synnyrsk massif is 300—350 million years. An age of 200 million years has also been established for this belt. The belt of Vitim alkaline rocks also conforms with the strike of structures in the Sayan-Baikal region and is situated north-east of northern Baikal. The aforesaid belt also includes alkali-gabbroid and, to a lesser extent, alkali-ultrabasic formations of the Saizhen complex belonging to two age groups: 350—220 and 230—170 million years.

The following patterns are observed in the distribution of alkaline massifs. The alkaline rocks of Prisayan are the oldest (650—550 million years) and the Jidinsk massif situated in the extreme south of the Sayan-Baikal mountain range is the youngest (119—110 million years). Accordingly, migration of alkali magmatism in different periods has been traced both along the direction of the Sayan-Baikal mountain belt from south-west to north-east as well as across its strike, from north to south.

The other feature of alkali magmatism is its confinement to specific linear zones—the aforesaid belts, representing zones of prolonged development of endogenetic stimulation of mantle. Herein, massifs of alkali-ultrabasic complexes are preferentially localised along these belts; at the same time the alkali-granitoid intrusive complexes (Aldan, Kunalei and other complexes) are developed on a considerably larger scale along the entire Sayan-Baikal tectonomagmatic activisation regions.

A few significant features are distinguished in the prerift epoch of the Baikal rift zone on the basis of the aforesaid. Magmatism generated at the mantle level was repeatedly manifested during the Palaeozoic and Mesozoic, but activisation processes were observed also during earlier epochs of development. Thus, according to A.A. Bukharov and E.P. Vasil'ev and others, narrow rift-like Olokit-type troughs were formed in the Northern

Pribaikal during the late Riphean and were filled with detrital facies rocks and volcanites of bimodal series having similar strikes with Cenozoic rifts, while extrageosynclinal-type volcanic belts formed in the middle Proterozoic. Thus, it may be presumed that a vast area of the mantle under the Baikal rift zone existed under a thermally stimulated state in the course of at least a few geological periods. All this attests to the high thermal activity of the Baikal rift zone during its prerift development epoch.

A similar situation has been established for the Hubsugul rift (northern part of the Mongolian People's Republic) belonging to the same rift system. According to the data of A.V. Il'in, formation of the late Riphean Darhat-Hubsugul rift with a submeridional strike preceded the Hubsugul (168) lake rift of Cenozoic age. The Cenozoic Hubsugul rift inherited its strike. Formation of the Darhat-Hubsugul rift of late Riphean age was accompanied by an injection of alkaline rock intrusives, differentiated gabbroids and volcanite flows of bimodal series. This region served as the arena for alkali magmatism (potash granites, syenites, nepheline syenites and urtites) in the Palaeozoic. Subalkaline rare metal-bearing granites were injected along old fissures during the Mesozoic. Thus, manifestation of a stimulated endogenetic regime of the mantle was recorded twice in the prehistory of the Hubsugul rift: in the late Riphean and in the Palaeozoic.

Activisation of Cenozoic alkali-basaltoid magmatism in the Dahomey-Nigeria belt was forestalled by the appearance of ring intrusives of alkali-granitoid complexes, situated in the submeridional belt stretching over 1500 km with a width of up to 200 km. The age of the intrusives varies from Jurassic (160 million years) in the south (Nigeria, Niger) up to middle Palaeozoic (285 million years) in northern Ahagar (Air). The alkali-granitoid massifs observed along the Cameroon lineament are of Tertiary age. Consequently, extrageosynclinal Cenozoic rift-type activisation in the Dahomey-Nigeria belt preceded alkali-granitoid magmatism of 'crustal' level magma generation. E.E. Milanovskii concluded on the basis of this belt that alkali-granitoid magmatism is characteristic for arch-block uplift-ment and is not typical for continental rift-forming zones. According to the present author, it is contemporary but not a specific type of magmatism for rift-forming regions.

The patterns of rift genesis, as revealed in the Red Sea-Aden rift zone, are also characteristic for other rift zones of East Africa. In this respect, the Rukva-Tanganyika rift zone coinciding with the Ubendi branch of the Mozambique belt is more visual.

The Ubendi zone underwent detailed complex history in the course of which its deeper parts were stimulated to a considerable extent and thermally 'shaken' by processes of multiple activisation, manifested in lower Proterozoic, Riphean and continued up to the end of the Precambrian, and was accompanied by alkali-ultrabasic, carbonatite, basaltoid and alkali-granitoid magmatism generated at varying depth. The Cretaceous and Cenozoic rift genesis in the Rukva-Tanganyika zone distinctly succeeded the stimulation sources in earlier epochs of Ubendi zone activisation.

It is important to emphasise that distinct inheritance is also clearly manifested in the location of the northern end of the Tanganyikan rift zone (Western rift). According to data from Kh. Belon and A. Poukle, prolonged (from late Cambrian up to the Paleogene) manifestation of prerift alkaline intrusive magmatism preceded the Cenozoic alkaline volcanism in the southern Kivu and Virung; this volcanism started considerably earlier (49 million years) than was priorly thought. Thus, the Luesh carbonatite (169) complex formed on the verge of 516 million years, the Virung nepheline syenites on the verge of 259 million years and the quartz porphyries and riebeckite granites in southern Kivu (Kahuzi) on the verge of 134 and 55 million years respectively. Furthermore, rift structures distinctly coincide with zones of maximum manifestation of Pan-African 'rejuvenation' and, consequently, with sources of earlier stimulation of the mantle [29].

Examples of the presence of prerift endogenetic regimes may be suggested in continental rift-forming zones. Let us, however, confine ourselves only to the Saint Lawrentian rift (Canada). Formation of this rift during the Cretaceous and Paleogene was preceded by alkali and basic magmatism, developed in its intersection zone with the Ottawa graben. Magmatism detectable in the Monteridgian alkali province was manifested in four age ranges from late Riphean to Mesozoic (1000–820, 565, 450 and 110 million years). Accordingly, prolonged interrupted endogenetic stimulation of the mantle preceded the Saint Lawrentian rift.

It follows from analysis of formational orders and their associated endogenetic regimes that areas characterised by more heated and thermally stimulated state of the mantle in the prerift epoch because of its prolonged and chronologically repetitive pulsational heating are the most suitable for 'rift generation'. Evolution of the endogenetic regime of the Red Sea rift zone conforms, most of all, to such a geodynamic situation, since thermal 'instability' of the mantle below it was at least two to three times more than it was below the Baikal zone. Deep-seated alkaline and basaltoid magmatism have been established here with six age boundaries (see Fig. 35).

The author distinguished the following evolutionary types of endogenetic regimes in continental rift-forming zones with their formational orders, as for example: 1) full or progressively recurring (progressive with respect to destruction of continental crust, recurring: i.e., reverse regime of mantle stimulation)—Red Sea and Aden rifts; 2) reduced or incomplete—Baikal rift and 3) embryonal or initial—Dahomey-Nigerian rift belt [27].

It is clearly evident from a comparison of formational orders that activised regions from typical rift to non-rift types sharply change in the character of prerift endogenetic regimes. First of all, simplification of the endogenetic regime takes place (it becomes less deep) and magmatism ascends, as though it had separated from its mantle source. The endogenetic regime in the Red Sea-Aden rift zone consistently retained its energetic potential and the mantle 'did not succeed' in cooling down before the Cenozoic. Consequently, a progressively recurring endogenetic regime provided maximum preparation ('maturing') of the lithosphere for rift formation. Its prime feature is prolonged pulsational stimulation ('instability') of the mantle, manifested in repeated generation of deep-seated mantle-derived alkali-basic and ultrabasic magmas and crustal granitoid (170) magmas. In the author's opinion, a leading role in the class of endogenetic rift regimes is played by the progressively recurring endogenetic processes.

It should be noted in conclusion that delineation of rift-forming endogenetic regimes is confronted with many problems of rift genesis, in particular, the interrelationship of these regimes in space and time. A directed nature of manifestation of endogenetic regimes is, first of all, established for the Red Sea rift zone, the gist of which lies in the fact that products of the stimulated regimes are localised along the axis of anticlinal zones (Fig. 40), whereas the distribution range of alkali-granitoid magmatism at crustal level of generation is considerably wider. Such directed endogenetic activisation continues also into the essentially rifting stage in the course of graben and axial trough formation, where contemporary manifestations of the stimulated mantle regime such as high level of heat flow, seismic activity, hot metal-bearing brines, weakly differentiated tholeiitic volcanism and others are observed.

Directed development of endogenetic regimes in the prerift epoch and their natural transition in the rifting stage may be regarded as an indication of inherited development of Cenozoic rift formation. This conclusion apparently touches upon one of the global problems, namely, formation of new oceans, insofar as the same tendency towards intensification of endogenetic regimes has been distinctly traced in mid-oceanic ridges. Such conformity in manifestation of endogenetic regimes in continental and oceanic rift zones and the concept of continental, intracontinental and oceanic rifts as different stages of the same process of continental crust destruction make it possible to presume that formation of oceanic rifts is, in many respects, predetermined by ancient endogenetic regimes and their traces are partially preserved in continental marginal mobile belts.

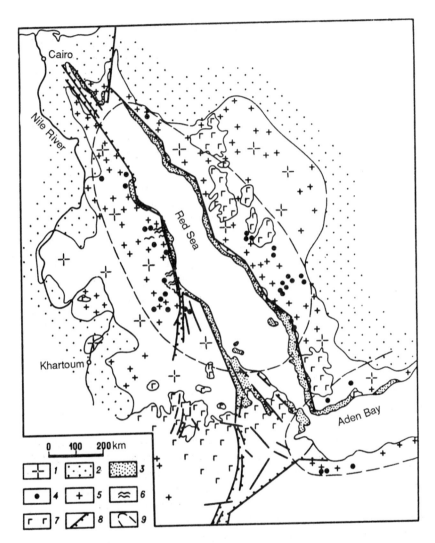

(171) Fig. 40 Scheme of disposition of basaltoid magmatism in the Red Sea-Aden rift zone.

1—Precambrian complexes of the Arabian-Nubian shield; 2—Phanerozoic platform cover; 3—Cenozoic deposits of rift depressions; 4—intrusive complexes of alkali and 'layered' gabbro; 5—intrusions of alkali-granitoid formations; 6—dolerite and diabase dykes; 7—Cenozoic basalts; 8—foremost rift-forming faults; 9—area showing development of basaltoid magmatism.

Correlation between Alkali Magmatism and Continental Rifting

Alkali magmatism occupies a specific place in the problem of predetermination of rift genesis. The confinement of rifting to regions with prolonged manifestation of alkali magmatism is well known and had been repeatedly emphasised by researchers. It is more clearly displayed in the East African rift belt, where spatial coincidence of alkaline and alkali-ultrabasic carbonatite magmatism with rift structures is particularly obvious (Western, Rukva-Nyassa, Kenyan and other rifts). However, this relationship is mainly accepted on a purely empirical basis.

(171) Formational analysis of the prerift epoch enabled a new assessment of the role of alkali magmatism in continental rift formation and correction of traditional concepts. It is known that alkali magmatism precedes and accompanies rift formation. However, this relationship is not universal and Cenozoic (and possibly more ancient) rift formation is not directly dependent upon the dimensions of its preceding alkali magmatism. For example, dimensions of alkali magmatism in the Red Sea rift zone distinctly fall short of those in the East African rifts, while Cenozoic rift formation had been stronger here both in dimensions as well as in depth of lithospheric transformation. This also concerns the Baikal rift zone, where the role of

(172) alkali magmatism has also been limited. These examples suffice to prove that alkali magmatism was characteristic but was not a definite factor in the history of prerift development of the continental rift-forming zones.

The relationship between alkali magmatism and rift formation was theoretically first suggested for the Kenyan rift by L. Williams in 1972, and later developed on a more scientific and factual basis by N.A. Logachev [19]. He showed from a lithothermal viewpoint that Cenozoic alkali magmatism is not confined to the entire rift system, but has a strictly defined location both in chronological (vertical) succession in rift formation and in space (lateral).

The patterns of manifestation of alkali magmatism in the Kenyan rift zone have been adequately detailed in the manuals of the Soviet East African Expedition as well as in the famous publications of N.A. Logachev and others. The Kenyan rift is one of the unique places linking Cenozoic alkali magmatism with rift formation. Attention should focus on the tectonic position of alkali magmatism manifested much before Cenozoic rift formation and only spatially coinciding with rift structures.

It has been shown on the basis of studies of the spatial-chronologic features of the prerift alkali magmatism of East Africa that they are often chronologically served from Cenozoic or even Mesozoic rift genesis. In this regard, the concept of the absence of correlation of alkali magmatism with rift genesis appeared at first glance not to be unfounded. For example, it is difficult to explain from the position of direct relationship of alkali magmatism with rift genesis why rifts accompany all the manifestations of alkali magmatism. The question posed by L.S. Borodin on how to explain the absence of ubiquitous control of alkali petrogenesis in the universal rift system appears justified, since it seems logical to expect such a control by the largest faults, namely of deep-seated mantle-originated melts.

New data on alkali magmatism significantly modify the concepts regarding its distribution and structural control. The data presented in Chapter 2 convincingly show that alkali magmatism is also characteristic of the Red Sea rift zone and the concept of its spatial relationship with rifts is likewise convincing. But the genetic aspect of this problem needs confirmation. Alkali magmatism in the Red Sea, Aden, Baikal, and other rift zones has no factual simple correlation with rifts. In almost all cases it has been established in the prerift development stage of these zones. Herein, this manifestation is, as a rule, significantly separated chronologically from the essentially rifting (graben forming) stage. L.S. Borodin has rightfully observed that alkali magmatism is not typical for rift genesis, if we take into account only the last stage of this process, which is (173) characterised mainly by considerably intensive stretching, i.e., entering far into the continental destructive process. Magmatism in this rift genetic process, as a rule, takes place under high energy potential, capable of melting only weakly differentiated alkali basaltoid and tholeiitic magmas. In other words, geodynamic and endogenetic energised conditions in an essentially rifting stage do not favour the appearance of deep-seated alkali-enriched magmatic melts.

Given the above-mentioned facts, the question arises of whether alkali magmatism may be generally related to rift formation. L.S. Borodin poses a question regarding the nature of contiguity of alkaline provinces of Eastern Africa with analogous rift systems in his book *Glavneishie provintsin i formatsii shchelochnykh porcd* [Main Provinces and Formations of Alkaline Rocks]. However, specific difficulties arise in direct correlation, insofar as rifts are of Cenozoic and, at least, Mesozoic age while alkali magmatism is, on the whole, more ancient. Thus, alkali carbonatite complexes in the Rukva-Tanganyikan segment of the Western rift is Upper Proterozoic-Vendian in age. As observed earlier, alkali magmatism in the northern segment of this rift was manifested in the Cambrian, Permian, early Cretaceous and Palaeocene. Alkaline rocks in the Red Sea rift zone are much older, being late Precambrian in age. According to the author, this discrepancy is surmountable if manifestation of alkali magmatism is linked not with contemporary rifts, but with prerift epochs of their development. Following this tectonomagmatic model, alkali magmatism localised in narrow linear zones (with a few exceptions) and spatial coincidence with Cenozoic rifts is regarded as the early stimulation process of the mantle along zones in which Cenozoic or Mesozoic rifts were subsequently formed.

The next situation arises out of the knowledge of tectonic patterns of manifestation of alkali magmatism, particularly in taking into account new data on the Red Sea rift zone. According to this situation, the commonly known pattern of confinement of alkali magmatism to the marginal parts of old platforms and folded regions or to the boundary zones of platform and folded regions needs confirmation. If alkaline provinces of East African provinces actually extend to marginal parts of the continent, then their position in the Red Sea zone is wholly controlled by other factors. As a rule, alkali intrusives are confined here to the uplifted zones of ancient basement or to the shoulders of 'rifts' and within the latter extend predominantly into deep trench-forming zones showing prolonged development.

Absence of direct relationship of alkali magmatism in the form of ring complexes with Cenozoic rifts has led researchers either to total negation of such a link or to attempts to associate it with other processes. Thus the publication of P. Gryutar and T. Vogel [46], dealing with the origin of alkaline ring complexes of Egypt, concluded on the absence of their direct link with rift or arch formation. It is suggested that they formed (174) from alkaline melts generated in the asthenosphere as a result of heating due to shear caused by movement of the African lithospheric plate. The authors rightly noted that alkaline melts intruded along activised Pan-African faults or along earlier existing weak zones. However, they were not in a position to explain the patterns of distribution, period of manifestation and tectonic control of ring intrusives. Ring intrusives in Egypt, as well as in the entire Red Sea framework, do not, on the whole, actually reveal a pattern at first glance, other than confinement to NE-striking faults and to intersection zones of the latter with NW-striking faults. As shown earlier, this feature of structural localisation constitutes only a part of the wider control of depth of ring intrusives.

Based on theoretical concepts of tectonic plates, envisaging movement of the African lithospheric plate along the asthenosphere, these researchers delineated six alkali magmation boundaries in Egypt (554, 404, 315, 229, 145–132, 91–89 million years) which are synchronous with known time intervals, marking changes in the movement of lithospheric plates. However, they did not proceed further in their attempts to link alkali magmatism with the movement of plates insofar as data on age of ring intrusives contradict the classical concepts of such a relationship. It is to be noted according to this concept that the manifestation of alkali magmatism in the form of ring intrusives is considered in some publications an indicator of plate movement until the separation of Africa and south America took place. Ring intrusives are regarded as traces in plates moving above the 'hot point' in the mantle. The Jurassic ring intrusives of the alkaline provinces in Nigeria, Angola, Damaraland and Nuanetsi, confined to NE-striking faults, are considered as traces of such 'hot points' in the mantle. Regeneration of massifs took place in the SW direction along these faults.

It is regarded in conformity with the above concept that Africa shifted in the north-east direction in the Jurassic and the North American plate moved towards the north-east direction. However, recent data on age and spatial distribution of ring intrusives of the Red Sea rift zone in Egypt and Sudan contradict the aforesaid concepts. Firstly, if the existence of a monolithic African lithospheric plate is accepted, which many researchers believe to have undergone displacement in the Palaeozoic and Meso-zoic in one direction, then a unidirectional lateral tendency of changes in age of ring intrusives can be expected; however, this is not actually observed within the entire Red Sea rift zone nor in individual lineaments. The South Egyptian fault (lineament) is very obvious in this respect; a chain of ring intrusives is 'threaded' on it from north-east to south-west, namely, Nugrus-El Tahtani (140 million years), Nugrus-El Fokani (139 million years), Mishbeh (142 million years), El Naga (145 million years), (175) El Hezira (229 million years) and Mansuri (132 million years). It is not difficult to perceive that a unidirectional chronological evolutionary ten-dency of alkaline complexes is absent in this lineament. If the Mansuri intrusive were excluded from it, then the sequential older age of intru-sives in the south-western direction could be explained from the point of view of 'hot points' with the assumption that Africa moved in a south-east direction. This concept contradicts data on adjacent intrusions and the above-mentioned lineaments, namely Nigeria, Angola, Damaraland etc.

Thus, it follows from the concepts presented in the present publica-tion that alkali magmatism spatially coincides with rift-forming regions, is characteristic for its prerift epoch and is genetically related to early stages of stimulation of the mantle. Consequently, spatial distribution of rift gen-esis and alkali magmatism belts, although somewhat separated by time, reflects their genetic relationship. Alkali magmatism precedes formation of rifts and continues during their development; it is a characteristic feature but not a determining factor for the formation of rifts. Geodynamic condi-tions of its formation led to no significant heating and thermal 'instability' of the lithosphere. The energy potential of the interior happened to be low in this case. Alkali magmatism 'works' more actively during multiple alternations with basaltoids of shallow depth, i.e., under a progressively recurring type of endogenetic regime.

Correlation of Endogenetic Regimes of Prerift and Rifting Stages of Development and Principle of Inheritance in Continental Rifting

Two contradictory but interrelated tendencies are displayed in the evo-lution of tectonic structures—their inheritance and superimposition [25]. Therefore, determination of inheritance envisages detection of quantita-tive correlation among these tendencies. The principles of inheritance are traditionally understood in geology as correlation among young and ancient tectonic structures according to their degree of similarity, disposition in space, dimensions, directional features and intensity of movements. Correlation of the newer structural plan of the Red Sea rift zone together with the prerift plan enables expansion of the scope for understanding the principles of inheritance and reviewing this zone in a broader perspective.

In terms of structure, detection of inheritance leads to establishment of the degree of conformity of fold and fault structures formed due to rifting together with the anisotropy of the substratum ('frame') for rift genesis. This simplest case reflects inheritance or superimposition of structures at the crustal level. In this respect, the Red Sea rift zone on the whole represents a superimposed structure. Herein, the degree of its superimposition sharply changes along the strike. Maximum superimposition (discordance) is manifested in the central segment, where the Red Sea rift intersects the Precambrian structures of the Sudanese-Arabian fold belt (176) at a large angle. Rift formation is particularly more intensive in this part of the Red Sea rift. Deep-water trenches with hot metalliferous solutions are located here and epicentres of earthquakes are concentrated in this region, i.e., endogenetic regimes are maximally manifested here. Such correlation seems paradoxical at first glance. But this becomes clear if we consider that the Red Sea rift followed the prerift endogenetic regime more fully predominantly at its central part. Such a relation can be regarded as an indicator of the interrelation of the rift and its substratum at different levels, i.e., the mantle and the crust albeit deviations are possible between them. The north-west structural trend is the determining factor in the case of the Red Sea rift and the same built up during the entire prerift epoch of mantle stimulation, i.e., the deep-seated mantle-based process, which formed the sharply superimposed and, to a certain extent, independent rift structures of the Red Sea, 'overcame', so to speak, the lesser depth (crustal) anisotropy of the 'framework' (substratum). Consequently, it is necessary to differentiate inheritance at different levels, i.e., in the mantle and in the crust in the course of development of rifts and, possibly, in other tectogenetic processes. Herein, the degree of correlation between inheritance and superimposition at these levels may change sharply—from inherited at the mantle level to superimposed (discordant) at the crustal, as in the case of the Red Sea rift. This conclusion conforms well with the concepts of A.V. Peive [25] regarding the development process of specific tectonic forms, such as: 1) major deep-seated tectonic structure and 2) minor tectonic developments on the surface should be differentiated. The prerift activisation 'cores', distinguished by the author, correspond to the major deep-seated tectonic structures of A.V. Peive, who conceived prolonged development through inheritance as characteristic for such structures.

Separation of prerift and essentially rifting epochs and the nature of their correlation in continental rift-forming regions show that the theoretical concepts regarding principles of inheritance in tectonics, as developed by N.S. Shatskii, A.V. Peive, A.L.Yanshin, V.E. Khain and other scientists, are also applicable to rift formation. A.V. Peive observes that three aspects of inheritance might be delineated in the case of tectogenetic problems, such as 1) tectonic plan, 2) tectonic form and 3) tectonic movements. From the viewpoint of separation of the prerift epoch and its correlation with the rifting epoch at the mantle and crustal (structural) level, the present author suggests that these basic indicators could be supplemented by a fourth, namely, inheritance at the level of endogenetic regimes.

The inheritance principles applicable to rift genesis are closely related to problems on the nature of boundaries of the prerift and rifting epochs of development in the continental rift-forming zones. As shown earlier, similarity of endogenetic regimes is characteristic of the prerift and essentially rifting stages and is revealed in the generation of basaltoid magmas alternating with alkaline magmas. Consequently, the endogenetic regime of continental rifts represents, as it were, 'penetrative' phases. Cyclicity of the endogenetic regime increased on transition from the prerift to the rifting epoch, whereas the duration of each cycle of mantle stimulation decreased with the number of such cycles increasing, which lead to intensification of the tectonomagmatic process. This feature is particularly obvious in its manifestation in the Red Sea and the Baikal rift zones. More than nine tectonomagmatic activisation epochs are distinguishable in the Red Sea rift zone (see Fig. 35). The duration of relatively clam epochs separating the tectonomagmatic activisation epochs, ranging from the Vendian to the Cenozoic, successively shortened, on average from 100 million years to 50–60, 40, and 30 million years during the Vendian-early Palaeozoic, middle Palaeozoic, Mesozoic and Cenozoic respectively. This pattern is more distinctly revealed in the Baikal rift zone. According to data presented by S.M. Zamaraev and others [31], the duration of tectonomagmatic activisation epochs was ~ 100 million years for the Palaeozoic, 70 million years for the early Mesozoic, 60 million years for the late Palaeozoic and 30–40 million years for the Cenozoic epochs. The duration of tectonically calm epochs changed accordingly. The periodicity and duration of individual tectonomagmatic activisation epochs are, on the whole, commensurable (although the number of such epochs is greater in the Red Sea zone) and a general tendency towards acceleration of tectonic processes has been preserved. Hence the repetition of tectonomagmatic cycles taking place in the development of the prerift endogenetic regime and the tectogenesis of an essentially rifting stage represent, as it were, the last link of these impulses, constituting thereby the logical conclusion in the development of a prerift endogenetic regime. In the light of such concepts, a distinct boundary between prerift and rifting epochs in the course of evolution of the endogenetic regime cannot be established.

If the type of magmatism and essentially endogenetic regime constitute, as it were, 'penetrative' phases, then the tectonic movements, their intensity and their trend in the rifting epoch qualitatively, but more so quantitatively, differ from the prerift epoch. This new geodynamic situation is characterised by a sharp increase in tectonic movements and growth of arch upliftments which, in some cases, preceded rift formation but in other instances, commenced with faulting accompained by vertical movement and graben formation without significant arch formation. However, a general feature of this qualitatively new development in the rift-forming zone is extensive shearing. In the case of deeper horizons, a similar situation was caused by the intrusion of the mantle diapir into the upper lithospheric horizons, followed by its movement and subsequent spread ('collapse'). This corresponds to the initial (shearing) stages of the mantle diapir [17] in the geodynamics of formation of the deep-water rift-generated trenches of marginal seas.

Thus we are confronted on the one hand by inherited development (magmatism and its types) and, on the other, by regenerated development (tectonic movements, discordance in structural plans) during the transition from prerift to rifting stages. In this regard, N.S. Shatskii's comment, emphasised by A.V. Peive, that there are neither 'pure' inherited nor 'pure' (178) superimposed structures seems to be realistic. Accordingly, both inheritance of endogenetic regimes as well as superimposition or regeneration of tectonic movements are distinctly manifested in the development process of the Red Sea rift formation zone. This fact underscores once again the complexity and contradiction in continental rift formation. In the opinion of A.V. Razvalyaev, the contradictory nature of continental rift formation, namely, its autonomy combined with predetermination, is revealed in the aforesaid situation.

It may be noted that the components of the inheritance or predetermination aspect, although still not formulated in a well-constructed concept, are described in publications by Soviet and other geologists. They are discussed in particular by V.E. Khain, E.E. Milanovskii and N.A. Bozhko. Based on the superimpositional and intersecting character of continental rift formation, A.F. Grachev reached the conclusion that rift formation is not related to preceding tectonic regimes, but, at the same time, he noted that a more detailed comparison of contemporary and ancient structures reveals some predetermination in laying of rift zones, which, possibly, causes specific features of magmatism [6]. He also remarked that in the opinion of some geologists, rift zones 'prefer' their formation either along ancient eugeosynclinal belts or along zones of prolonged manifestation of alkali magmatism. Thus, he has not denied the deeper link of rift formation with preceding history. He noted rightly that, if the predetermination phenomenon has been manifested roughly in the African-Arabian rift belt and the Baikal rift zone, then it is difficult to trace in the Mediterranean Sea-Moss Lake (Western Europe) rift generated zones, where the latter intersects (reveals discordancy) the structural pattern of the basement. Actually, such is the case but, as will be shown, this type of contradiction cannot serve as testimony against the predetermination principle in rift formation. Let us note that the Red Sea rift zone is confined, on the whole, to the late Proterozoic fold belt and has developed more intensively precisely in those regions where it is maximally discordant to its individual fold branches (Sudanese-Arabian). Hence the fact of structural non-coincidence does not signify the absence of inheritance and predetermination in the development of the rift-formation process. The predetermination problem in rift genesis should be solved by generalised analysis of data on constitution of rift-formed zones and, first of all, evolution of endogenetic regimes manifested in magmatism.

V.S. Fyfe, O.K. Leonardos, D. Bailey and R. McConnell are among those who have discussed the components of inheritance or the predetermination aspect in rift formation. D. Bailey's concept [15] on rift formation and the association of alkali magmatism with it is based on deformation and fragmentation of the rigid plates ('frame') accompanied with the fall of pressure and, as a result, melting of sublithospheric parts (upper mantle) with escape of volatile components enriched with alkalis. The form of the migration route of volatiles or heat flows (heat transfer)—whether along points, linear or along the surface—determines the nature of manifestation of magmatism (volcanoes, central type intrusion, lava plateau). According to D. Bailey, rifts are originated in this way and subsequently develop spontaneously, being supported by the ascent of heat and volatiles from the mantle and appear to be autonomous in this respect. Moreover, attention is paid in the aforesaid work to the energetic state of one or the other territory until the beginning of arch formation. Regions where prerift magmatism is weakly manifested or is totally absent were characterised by exclusively low heat flow during the period preceding Cenozoic rift formation. D. Bailey observes that melting of the upper mantle or lower lithosphere due to lower pressure under such conditions was perhaps not possible because of low energy potential, i.e., character of rift formation and degree of its advancement depend upon the preceding energy state of the deep-seated interior. Examples of correlation of rifts with the basement anisotropy have been discussed by F. Dicksie, R. McConnell and others. However, none of these authors could reach a well-conceived idea regarding the inherited development of rift formation, particularly at the level of endogenetic regimes.

It is necessary to differentiate predetermination in the pattern of inherited development of a rifting epoch from the prerift endogenetic regime and also the structural plan caused by anisotropy of the substratum of rift zones in the case of continental rifting. These phenomena need not necessarily fully coincide. The dependence of rift formation upon structural anisotropy of the substratum is well known and has been sufficiently elaborated. Detection of a deeper link is important for us. In the author's opinion, the latter constitutes evolution of deep-seated endogenetic regimes preceding rift genesis. If the mantle was not sufficiently heated at the moment of rift origin (in cold energetic state), then its manifestation is reflected only in arch formation, as for example, in many regions of Africa (Dahomey-Nigerian belt, Tibesti, Haruz in Libya, Western Sudan and others) and in the appearance of alkaline basalts (trachytes, phonolites) and ultra-alkaline silicic rocks such as comendites. If rift formation was preceded by an adequately heated and stimulated state of the mantle and, in an optimal case, by a progressively recurring endogenetic regime, it continued vigorously, accompanied by arch formation eruption of weakly differentiated tholeiitic basalts in lesser quantity, graben formation, thinning of granite-metamorphic layer and, finally, its total severance and intrusion of mantle material (Red Sea, Aden Bay).

Possibly, the Cenozoic rift formation (or rift formation as a whole) is more distinctively developed according to the given model mainly in regions of ancient Precambrian consolidation, which provided maximum stability to the lithosphere.

Rift formation starts with upliftment and arch formation. This stage may differ in duration and dimensions of manifestation. Apparently, it will be prolonged and, possibly, more distinctly revealed in structures (180) formed in a 'colder' mantle with low energy gradient (Tibesti, Hoggar and others). Active arch formation reduces in rifts where it is preceded by a stimulated regime, since the rift-formation process takes place at a fast rate and the dominant role in its development is played by a graben-formation stage and less by alkali magmatism. For example, the Red Sea and Aden rifts are characterised by vast but relatively weakly manifested arches in the prerift stage and more basic volcanism, while the Kenyan rift has a well-defined arch and more intensive alkali magmatism. The cited examples illustrate this situation but not fully insofar as the Red Sea, Aden and Kenyan rifts were formed within the vast Red Sea-Mozambique belt of tectonomagmatic activisation and the Kenyan rift was preceded by, though to a lesser extent, an adequately formed and stimulated regime of the mantle.

The starting point of rift formation is heating of the mantle. D. Bailey and S.I. Sherman consider that this process may be initiated either by selective melting in the fault zone during shearing, or by global thermal stimulation of the mantle, as for example, in the Meso-Cenozoic. In either case, the rigid lithosphere reacts to the above change first at places of its anisotropy (fault, deep-seated flexure, fold belt and so on). Further, heating leads to the appearance of a layer with an anomalously heated, thinned mantle, favouring arch formation and, subsequently, together with the progress of the rift-forming process, its shearing (intense graben formation) with partial or full severance of the granite-metamorphic layer of the crust. Such is the main pattern of rift formation. To what extent this can be achieved depends upon the type of prerift endogenetic regime preceding rift formation.

Activisation of the mantle during the rifting regime is localised along specific linear zones with increased permeability. It is known that the development of Cenozoic rift formation in the continents was controlled by tectonomagmatic activisation zones. However, this association was too general and reflects a global pattern, i.e., confinement of rift formation to regions with increased thermal stimulation and high energy potential. Rifts display a higher order of selectivity even within the tectonomagmatic belts. A link of Cenozoic rift formation has been established with specific structures characterised by prolonged and consistent thermal stimulation of the mantle in the Red Sea-Aden and other rift zones for example. Exclusive linearity of these zones within the boundaries of tectonomagmatically active zones and their occurrence over considerable areas lead us to distinguish these structures as 'cores' of activisation or 'rift-formed cores'. The latter are represented by zones of maximum manifestation of an endogenetic regime. The general faults distinguished by S.I. Sherman for the Baikal rift may be indications of such 'cores' that might have initi-
(181) ated rift formation or the initial faults or 'incisions' which, in the opinion of V.V. Belousov, trace the rift depression and precede its formation.

Following the concept of the predetermination of rift formation developed by the author 'rift cores' represent zones of prolonged, periodic stimulation of the mantle in the prerift development epoch. These activisation zones ('cores') proved more matured for rift-forming processes during the Cenozoic period, characterising global activisation, and became places with emergence of mantle diapirs, thereby predetermining the linearity of the latter. Global tectonic activisation is, first of all, observed in those places where similar endogenetic regimes already prevailed, i.e., in regions with a heated 'stimulated' mantle. Consequently, global activisation epochs, whose existence has again been confirmed [18] in recent years, only accelerated the evolution of endogenetic regimes progressing in one or the other rift-forming region, i.e., they may be regarded as a hastening ('accelerating') factor in rift formation. Rift genesis manifests itself precisely in prerift stimulation zones of the mantle, i.e., a deep-seated mantle-based process is the leading factor in rift formation.

A combination of data, including the development history of the Arabian-Nubian shield in the Precambrian and the distribution pattern of lateral-chronologic orders of magmatic formations, provides indicators of endogenetic regimes both for the prerift as well as for rift-formation epochs. Such indicators lead us to consider that the 'rift-forming core' of the Red Sea zone is a deep-seated structure of prolonged development, since its historical roots extend up to the early stages of formation of the earth's crust, apparently into its protometamorphic layer. According to Yu.M. Sheiman, structures of a similar type conform to the concept of 'tectonosphere'.

The development of the Red Sea rift zone may be represented in the form of a series of palaeotectonic profiles which depict the place and time of manifestation of rift formation and its relationship with the ancient structure (Fig. 41).

Geophysical data on the deep-seated constitution of the Red Sea rift zone tectonosphere are inadequate. However, investigations carried out in this direction still enable us to obtain some idea of the features of the deep-seated constitution of this territory. Observations made at the seismic station situated at Shiraz (Iran), Jerusalem (Israel), Heluan (Egypt) and Addis Ababa (Ethiopia) show that the upper mantle under Arabia is intermediate in constitution between the mantle below ancient shields and that below young stable regions. According to this data, the thickness of the crust in Arabia is estimated to be 35 ± 8 km. Approximately the same thickness has been derived for the African part of the Arabian-Nubian Shield. A layer of lower (seismic) velocities adjacent to the asthenosphere has been detected at 100—140 km depth in Arabia. The lower boundary of this layer is situated at approximately 220 km below the Mohorovičić surface. J. Fairhead and K. Riva established from (183) seismologic and gravimetric investigation data that the asthenospheric layer occurs at 175 km depth from the earth's surface due west of the Red Sea and gradually becomes buried down to a depth of 225 km and is still deeper towards the Central-African craton. Change in the depth of occurrence of the asthenospheric layer takes place in the Red Sea rift zone and coincides with the trench zone formed even in the Precambrian. Herein, the difference in its depth of occurrence is estimated to (184) be 50—70 km. Accordingly, emergence of anomaly in the mantle in the Red Sea rift at ~ 10 km depth may be logically linked with rupture of the mantle diapir at the boundary of two large geoblocks (Nubian and Arabian) with different constitution of the upper mantle. Fig. 42 depicts a hypothetical model of the deep-seated constitution of the Red Sea rift zone taking into account the cited geophysical data and historic-geologic development of the Red Sea rift zone. It should be noted that the given model of the Red Sea rift zone exhibits similarity with the Baikal rift zone; the structural position of the latter is due to its confinement to the junction of two very large lithospheric blocks—the Siberian platform and the heterogeneous Syano-Baikal fold region.

The patterns of rift genesis, as revealed in the Red Sea rift, are also characteristic for other continental rift zones. In this respect, the Baikal rift zone is more obvious; N.A. Bozhko, S.M. Zamaraev, V.A. Naumov [24] and others developed similar views regarding the nature of its rift formation (cyclicity, duration, inheritance of development from the preceding history). These data together with the aforesaid statements show that the correlational patterns of prerift and rifting structural set-up are characteristic for the two largest continental rifts of the planet and justification for the delineation of a prerift epoch in the development of continental rift formation can hardly be doubted. At the same time, we are far from the concept that development of all continental rift zones takes place following the Red Sea model. first of all, it should be emphasised that the established patterns are inherited by rift zones developed on ancient platforms which underwent prolonged tectonomagmatic activisation. Rift genesis of the mentioned type is usually related to tectonomagmatic activisation belts and tectonothermal transformation of Mozambique, Dahomey-Nigeria, Grenville, Stanovoi type and others. A large hiatus between tectonomagmatic events determining their specificity and subsequent activisation preceding rift formation is particularly absent in such belts. For example, the prerift activisation magmatism in the form of ring intrusives in the Red Sea rift zone followed the orogenic epoch with a minimum hiatus and continued up to the Cenozoic rift formation.

In conclusion, it should be stressed that it may not be possible to identify the prerift stage with the same degree of soundness in all rift zones. This fact refers particularly to young epipalaeozoic platforms, partially to marginal-continental or trans-oceanic rift-type submergences and, to a considerable extent, Precambrian green-stone-bearing trenches and palaeovolcanites. However, lack of any 'universality' in isolation of a prerift epoch cannot constitute an inadequacy in the mentioned approach to the study of continental rift formation, let alone an argument against justification of its delineation. E.E. Milanovskii considers that rift zones are varied (185) not only in intensity of magmatic phenomena, but also in prehistory, general set-up, structure of substratum, tectonic constitution, kinematic movement and history of formation [22]. Therefore attempts to force all varieties of rift zones into the 'procrastination bed' or into an 'all-enveloping' model appears to be methodically unreliable. This would extremely narrow down the boundaries of our knowledge of rift genesis and would unduly simplify all the diversities of these natural phenomena. Classification of rift

188

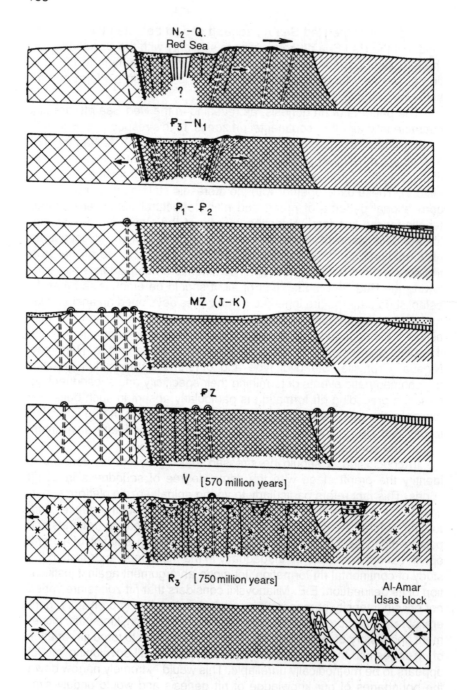

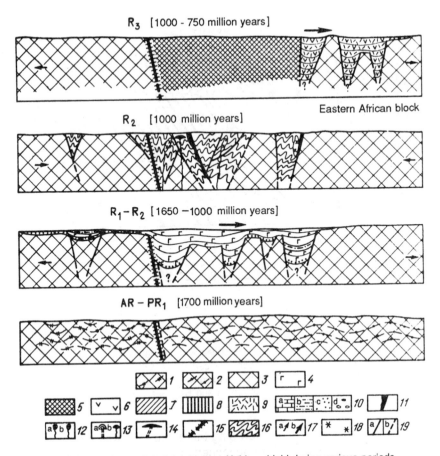

R₃ [1000 - 750 million years]

Eastern African block

R₂ [1000 million years]

R₁-R₂ [1650 −1000 million years]

AR − PR₁ [1700 million years]

82–183) Fig. 41. Palaeotectonic profile of the Arabian-Nubian shield during various periods.

1 to 3—early Precambrian continental crust: 1—Nubian block (gneisses, crystalline schists, marbles of granulitic and amphibolitic facies metamorphism), 2—Arabian block (gneisses, crystalline schists of amphibolitic facies metamorphism), 3—same, without mention of material composition; 4—early to middle Riphean volcanogenic-sedimentary complex of intracratonic troughs (basalts, andesites, tuffs, schists, greywackes, marbles); 5—crust consolidated towards end of middle Riphean; 6—late Riphean volcanogenic-sedimentary complex of intracratonic troughs (andesites, basalts, dacites, rhyolites, tuffs, sandstones, shales, marbles); 7—crust, consolidated towards end of late Riphean; 8—terrigenous-carbonate, continental and marine coastal deposits of Palaeozoic age; 9—acid volcanic rock; 10—continental and marine shallow-water sediments (a—limestones and dolomites, b—clays and sandstones, c—sandstones, clays and limestones, d—boulder-pebble conglomerates); 11—alpine type ultrabasites; 12—granites (a—Calc-alkaline, b—alkaline and subalkaline); 13—intrusions (a—ring, alkaline composition, b—'layered' gabbro); 14—basalt sheets and their feeder canals, 15—boundary between Arabian and Nubian metablocks (furrow zone); 16—folded complexes; 17—diections (a—tectonic stresses, b—lateral displacement during consolidation of crust); 18—regions of manifestation of Pan-African thermotectogenesis; 19—faults (a—proved, b—assumed).

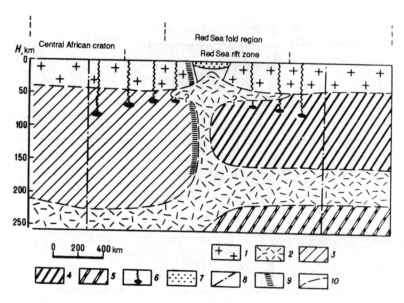

(185) Fig. 42. Hypothetical model of deep-seated constitution of Red Sea rift zone.

1—continental crust; 2—asthenosphere; 3—lithospheric part of the mantle, active during period of development of the Red Sea rifting phase (PR_2); 5—upper mantle, active during period of development of the Red Sea rifting phase; 6—magmatically active zones in prerift development epoch of the Red Sea rift zone; 7—sediments in rift depression; 8—assumed contours of tectonosphere, active during prerift development epoch; 9—furrow zone in lithosphere between the Central African craton and Red Sea fold region; 10—provisional boundary of tectonospheric layers.

zones by E.E. Milanovskii, sub-dividing them into fractured arch-volcanic epiplatform, epiorogenic and intracontinental rifts, may be supplemented by their inherited or predetermined historical geologic types.

Conclusion

(186) Analysis and generalisation of results arising out of the study of geological structure and developmental history of the northern part of the African-Arabian rift belt and other continental rift zones have generated the following conclusions:

1. The relationship of continental rift genesis with preceding geologic history has been theoretically founded with the unique examples of the Red Sea and other rift zones of the planet. A prerift epoch has been distinguished in its development, which signifies a combination of tectonomagmatic processes occurring in one or the other rift-forming region after formation of a mature continental crust in it prior to the structural-morphologic formulation of rift.

2. Continental rifts develop as endogenetically active systems. Mantle diapirism is the leading factor in their development; it leads to destruction of the continental crust and, eventually, to its replacement by regenerated oceanic crust. Source-based (from specific point or diffused) stimulation of the mantle, localised along linear zones, i.e., activisation 'cores', correspond to the prerift stage. Unification of the sources into linear projections of the asthenosphere took place in the preliminary rift stage of the rifting epoch and zones of higher permeability formed above them. The growth of asthenospheric (mantle based) diapirs and their flow (spreading) corresponds to the essentiality rifting stage.

3. The association of continental rift formation with the preceding geologic history is manifested in structural plan on the basis of dependence of rift formation upon the strikes of fault and fold structures, composition and degree of 'rigidity' i.e., ansotropy of the prerift substratum. A deeper link consists in the inherited development of endogenetic regimes of prerift and rifting epochs.

4. Continental rift formation developed more actively in those regions where the lithosphere, at the moment of its appearance in the prerift epoch, was more heated, 'thermally unstable', or destabilised. Parts of the earth's crust showing development of a progressively recurring endogenetic regime in the prerift stage display maximum rift formation; its major characteristics are represented by prolonged pulsational stimulation of the mantle, manifested in repeated generation of magmas at different depths. A similar situation favours 'maturing' of the lithosphere for rift

formation much before its main paroxysm in the Cenozoic.

5. The known contradiction in continental rift formation is exhibited on the one hand in its autonomy and, on the other, in the inheritance from preceding history; the same is explained by the correlation of structural patterns of the prerift and rifting epochs at different levels, i.e., the mantle and the crust. At the mantle level, predetermination is caused by succession of endogenetic regimes, and at the crustal level by dependence of rift formation upon the structural anisotropy of the basement.

(187)   6. The geological prerequisites for continental rift formation are determined by prolonged interrupted thermal stimulation of the mantle in the prerift epoch, as also by major lithospheric inhomogeneities, occurring even in the early Precambrian.

7. The delineated patterns of spatial and chronological distribution of magmatic complexes in the prerift epoch with specific metallogenic specialisation have great significance for regional and local forecasting of ore deposits. The patterns of structural localisation of manganese deposits in the Red Sea coast and deep-water basins with hot metal-bearing sediments are of special methodological significance; they serve as 'laboratories' of contemporary ore formation in oceans while studying other continental rift zones and, in particular, their ancient analogues.

Selected References

1. Al'mukhamedov A.I., G.D. Kashintsev and V.V. Matveenkov. Evolyutsiya Bazal'to-vogo Vulkanizma Krasnomorskogo regiona [Evolution of Basalt Volcanism in the Red Sea Region]. Nauka, Novosibirsk (1985).

2. Artyushkov E.V. Geodinamika [Geodynamics] Nauka, Moscow (1979).

3. Belousov V.V. Éndogennye rezhimy materikov [Endogenetic Regimes of Continents]. Nedra, Moscow (1978).

4. Bozhko N.A. Tektono-termal'naya pererabotka dokembriĭskogo fundamenta Gond-vani [Tectono-thermal transformation of the Precambrian basement of the Gondwana]. Vest. MGU, Ser. Geol., no.5, pp. 17–30 (1979).

5. Bozhko N.A. Pozdnii dokembrii Gondvani [Late Precambrians of the Gondwana]. Nedra, Moscow (1984).

6. Grachev. A.F. Riftovye zony Zemli [Rift zones of the Earth]. Nedra, Leningrad (1977).

7. Grachev A.F., Yu.S. Genshaft and A.Ya. Saltykovskii. Geodinamika Baikalo-Mon-gol'skogo regiona v kainozoe [Geodynamics of the Baikal-Mongolian region in the Cenozoic]. In: Kompleksnye issledovaniya razvitiya Baikalo-Mongol'skogo regiona v kainozoe. Moscow (1981), pp. 134–176.

8. Dolginov E.A., V.P. Ponikarov and A.V. Razvalyaev. Endogennye rezhimy v svyazi s problemoi kontinental'nogo riftogeneza. [Endogenetic regimes in relation to problems of continental rift formation]. In: Geologiya Al'pid "Tetisnogo" proiskhozndeniya. Tektonika, Moscow (1980), pp. 139–143.

9. Zonenshain L.P., A.S. Monin and O.G. Sorokhtin. Tektonika Krasnomorskogo rifta v raione 18°S. Sh. [Tectonics of the Red Sea Rift in the lat. 18°N]. Geotektonika, no. 2, pp. 3–22 (1981).

10. Ivanov, S.N. Ofioliti na sialicheskoi kore. [Ophiolites in the sialic crust]. In: Evolyut-siya ofiolitovykh kompleksov. Sverdlovsk (1981) pp. 72–78.

11. Isaev, E.N. Struktura zemnoi kory v zone sopryazheniya Krasnomorskikh gor i Krasnogo Morya v svete novykh geofizicheskikh dannykh [Structure of the earth's crust in zones of contiguity between the Red Sea hills and the Red Sea in the light of new geophysical data]. Bull. MOIP, Ser. Geol., 57, 1, 31–39 (1982).

12. Isaev E.N., V.V. Samoilyuk and N.A. Shabalin. Strukturno-geofizicheskaya model' stroeniya vulkanogennoge-osadochnogo chekhla i Fundamenta Adensko-Krasnomorskogo Regiona [Structural-geophysical model of constitution of the volcanogenic sedimentary cover and basement in the Aden-Red Sea region]. In: Tez. dokl. 27 MGK, Moscow, vol. 13, pp. 232–233 (1984).

13. Kaz'min V.G. Riftovye struktury Vostochnogo Afriki-raskol kontinenta i zarozhdenie okeana [Rift structures in Eastern Africa—Splitting of continents and Origin of Oceans]. Nauka, Moscow (1987).

14. Koneev A.A. Nefelinovye porody Sayano-Baikal'skoi gornoi obslasti [Nepheline Rocks in the Sayano-Baikal Hill Region]. Nauka, Novosibirsk (1982).

15. Kontinental'nye Rifty [Continental Rifts]. Edited by I.V. Ramberg and E.R. Neiman. Mir, Moscow (1981).

16. Lazarenkov V.G. Ovremennom i lateral'nom ryadakh shchelochnikh platformen-

194

nykh formatsii [On chronologic and lateral orders of alkaline platform formations]. *Geologiya i Geofizika*, no. 11, pp. 61–69 (1987).

17. Levin L.E. Geologiya okrainnikh i bnutrennykh morei [Geology of Marginal and Internal Seas]. Nedra, Moscow (1979).

18. Leonov Yu.G. Orogennye periody i epokhi tektogeneza kak formy proyavleniya global'noi tektonicheskoi aktivnosti. [Orogenic periods and tectogenetic epochs as forms of manifestation of global tectonic activity]. In: Geologiya Al'pid "Tetisnogo" Proiskhozhdeniya. Tektonika, Moscow (1980), pp. 149–158.

19. Logachev N.A. Vulkanogennye i osadochnye formatsii riftovykh zon vostochnoi Afriki [Volcanogenic and Sedimentary Formations in Rift Zones in Eastern Africa]. Nauka, Moscow (1977).

20. Logachev N.A., Yu.A. Zorin and S.I. Sherman Geodynamika kontinental'nykh riftov [Geodynamics of continental rifts]. *Geologiya i Geofizika*, no. 12, pp. 13–22 (1982).

21. Milanovskii E.E. K probleme proiskhozhdeniya i razvitiya lineinikh struktur platform [On problems of origin and development of linear structures in platforms]. *Vest. MGU, Ser. Geol.*, no. 6, pp. 29–58 (1979).

22. Milanovskii E.E. Riftovye zony kontinentov [Rift Zones of Continents]. Nedra, Moscow (1976).

23. Moskaleva V.N. Magmaticheskie formatsii kak indikatory riftogennykh sistem [Magmatic formations as indicators of rift-forming systems]. *Sov. Geologiya*, no. 10, pp. 82–93 (1982).

24. Naumov V.A. Geologicheskaya Predystoriya Baikal's-kogo rifta. [Geologic prehistory of the Baikal Rift]. In: Tektonika i seismichnost' kontinental'nykh riftovykh zon. Moscow (1978), pp. 47–61.

25. Peive A.V. Prinstsip unasledovannosti v tektonike [Principles of inheritance in tectonics]. *Izv. AN SSSR. ser. geol.* no. 6, pp. 11–18. (1956).

26. Razvalyaev A.V. K probleme pozdneproterozoiskikh giperbazitovykh poyasov Araviiskogo-Nubiiskogo shchita [On the problems of late Proterozoic ultrabasite belts of the Arabian-Nubian shield]. *Geotektonika*, no. 4, pp. 19–32 (1979).

27. Razvalyaev A.V. Endogennye rezhimy predshest vuyushchie riftogenezy [Endogenetic regimes preceding rift formation]. *Geotektonika*, no. 6, pp. 32–41 (1979).

28. Razvalyaev A.V. Doriftovyi etap razvitiya Krasnomorsko-Adenskoi zony [Pre-rift epoch of development of the Red Sea-Aden Zone]. *Geotektonika*, no. 1, pp. 85–98 (1984).

29. Dolginov E.A., I.V. Davidenko, A.V. Razvalyaev and N.A. Stikhotvortseva. Rudonosnye formatsii dokembriya Vostochnoi Afriki i Aravii [Ore-bearing Formations in the Precambrians of East Africa and Arabia]. Nedra, Moscow (1979).

30. Suludi-Kondrat'ev, E.D., A.V. Razvalyaev, I.V. Davindenko et al. Systemy razlomov Afriki i Azii [Systems of Faults in Africa and Arabia]. Nedra, Moscow (1984).

31. Zamaraev S.M., E.P. Vasil'ev, A.M. Mazukabzov et al. Sootnoshenie drevnei i kainozoiskoi struktury v Baikal'skoi riftovoi zone [Correlation of Ancient and Cenozoic Structures in the Baikal Rift zone]. Nauka, Novosibirsk (1979).

32. Khain V.E., Destruktivnyi takto genez i ego global'noe poyavlenie [Destructive tectogenesis and its global manifestation]. In: Tektonika i strukturnaya geologiya. Planetologiya (Internat. Geol. Congress, XXVI Session, Reports of Soviet Geologists), pp. 5–13. (1976).

33. Yashina R.M. Nefelin-sienitovyi magmatizm v kontinental'nykh strukturakh s razlichnym tektonicheskim rezhimom. [Syenitic Magmatism in the continental structures with different tectonic regimes]. In: petrologiya (Internat. Geol. Congress, XXVI Session, Reports of Soviet Geologists, Moscow), pp. 129–150 (1980).

34. Serensits, C.M., H. Faul, K.A. Foland *et al*. Alkaline ring complexes in Egypt: Their ages and relationship in time. *J. Geophys. Res.*, 86, 4, 3009–3013 (1981).

35. Almond, D.C. Precambrian events at Saboloka, near Khartoum and their significance in the chronology of the basement complex of North-East Africa. *Precambrian Res.*,

13, 1, 43–62 (1980).

36. Basahel, A.N., U. Jux and S. Omara. Age and structural setting of a Proto-Red Sea embayment. *N. Jb. Geol. und Paläntol. Monatsh*, no. 8, pp. 456–468 (1982).

37. Bermingham, P.M., J.D. Fairhead and G.W. Stuart. Gravity study of the Central African Rift System: A model of continental disruption. 2. The Darfur domal uplift and associated Cainozoic volcanism. *Tectonophysics*, vol. 94, pp. 205–222 (1983).

38. Cochran, I.R. A model for development of the Red Sea. *A.A.P.G.Bulletin*, 67, 1, 41–69 (1983).

39. Delfour, I. Geologic, tectonic and metallogenic evolution of the northern part of the Precambrian Arabian shield (Kingdom of Saudi Arabia). *Bull. Bur. Res. Geol. et Mineras*, sec. 2 nos. 1–2, pp. 1–19 (1980).

40. Ei Ramly, M.F. and A.A.A. Hussein. The ring complexes of the Eastern Desert of Egypt. *J. African Earth Sci.*, 3, 1/2, 17–39 (1985).

41. Klerkx, G. Evolution tectonometamorphique du Socle precambrien de la region d'Uweinat, Libye. *Rev. Geol. Dyn. Geograph. Phys. Sci.*, 21, 5, 319–324 (1979).

42. Kröner, A., M.I. Roobol, C.R. Ramsay and N.I. Jackson. Pan African ages of some gneissis rocks in the Saudi Arabian Shield. *J. Geol. Soc. London*, vol. 136, pp. 455–461 (1979).

43. Kröner, A. Ophiolites and the evolution of tectonic boundaries in the Proterozoic Arabian-Nubian Shield of North-East Africa and Arabia. *Precambrian Res.*, 27, 5, 277–300 (1983).

44. McConnell, R.B. A resurgent taphrogenic lineament of Precambrian origin in Eastern Africa. *J. Geol. Soc.*, 137, 4, 483–489 (1980).

45. Meinhold, K.D. The Precambrian Basement Complex of the Bayuda Desert, Northern Sudan. *Rev. Geol. Dynam. Geogr. Phys.*, 21, 5, 395–401 (1979).

46. Phil de Gruytar and T.A. Vogel. A model for the origin of the alkaline complexes of Egypt. *Nature*, vol. 291, pp. 571–574 (1981).

47. Stacey, J. S. and C.E. Hedge. Geochronologic and isotopic evidence for early Proterozoic crust in the eastern Arabian Shield. *Geol. Soc. Amer. Bull.*, vol. 96, pp. 817–826 (1984).

48. Stoeser, D.B. and V.E. Camp. Pan-African microplate accretion of the Arabian Shield. *Geol. Soc. Amer. Bull.*, vol. 96, pp. 817–826 (1985).

49. Vail, J.R. Pan-African (Late Precambrian) tectonic terrains and the reconstruction of the Arabian-Nubian Shield. *Geology*, 13, 12, 839–842 (1985).

50. Vail, J.B. Alkaline ring complexes in Sudan. *J. African Earth Sci.*, 3, 1/2, 51–59 (1985).

INDEX OF THE MOST IMPORTANT STRUCTURES[*]

---
[*]Original Russian page numbers are given here as well as in the left margin of the English translation.

Printed in India.